The Threat and the Glory

The Threat

and the Glory

Reflections on Science and Scientists

Peter Medawar

Edited by David Pyke

Foreword by Lewis Thomas

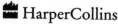 HarperCollins

A Cornelia & Michael Bessie Book
An Imprint of HarperCollins*Publishers*

FIRST U.S. EDITION

Library of Congress Cataloging-in-Publication Data

Medawar, P.B. (Peter Brian), 1915-
    The threat and the glory : reflections on science and scientists /
  P.B. Medawar ; foreword by Lewis Thomas ; edited by David Pyke.
      p. cm.
    "Originally published in Great Britain in 1959"—T.p. verso.
    ISBN 0-06-039112-X
    1. Science—Philosophy.   2. Science—Moral and ethical aspects.
  3. Scientists—Moral and ethical aspects.   I. Pyke, David.
  II. Title.
  Q175.M435   1990   89-46107
  501—dc20

90 91 92 93 94 HC 10 9 8 7 6 5 4 3 2 1

It is the great glory as it is also the great threat of science that everything which is in principle possible can be done if the intention to do it is sufficiently resolute. Scientists may exult in the glory, but in the middle of the twentieth century the reaction of ordinary people is more often to cower at the threat.

'*The Threat and the Glory*', *Chapter 2*

Foreword

Lewis Thomas

Peter Medawar possessed more friends, all round the world, than anyone I have ever known or heard of. Possessed is the word: they hung on his words, read with close attention and vast pleasure everything he wrote, rejoiced in his achievements, worried endlessly about his health, wondered at his knack for survival and his outright defiance of all the rules of neurology, and adored him. Fifty, a hundred years from now, Medawar will be a steady source of Ph.D. theses for graduate students from any number of disciplines—biology and immunology, of course, since he and his colleagues opened up the fields cleared by Ehrlich, Bordet, and Landsteiner in earlier generations, revealing in a stroke of classical Medawar experimentation the new and entrancing depths lying just below. I can already imagine the doctoral dissertations: the tolerance of mice, the tolerance of twin cows, something less than tolerance of psychoanalysis, outright immunologic rejection of Teilhard de Chardin, the acceptance and revascularization of Francis Bacon.

Then there will be the other scholars, deep in another wing of the library stacks, the Eng. Lit. graduate students on the track of Medawar the twentieth-century man of letters. His books—those by Peter and those others written by Peter and his wife Jean—will still be on the shelves, together with any number of proceedings of international congresses and biological symposia. These latter will, I predict, be of special interest, for Peter is in those proceedings, on the record, with some of the wisest and wittiest remarks of the twentieth century.

Something that I hope will not be missed: the electromagnetic energy of this spectacular man, his exuberance, his outrageous disrespect for so many canons (of all stripes and definitions), and most important of all, his own deep pleasure in life, his immense respect for Jean, and all the fun he had all his life, with his astonishing and capacious mind.

And then, of course, there will be the philosophy students, fingering earnestly through the stacks trying to make sense of the twentieth century and discovering, I hope, through a close reading of Medawar's works, that the century made whatever sense it turned out to make because of science. This was his central conviction, the source of his faith and pleasure, his tenacious hope for the future.

He has left intellectual marks all through the laboratories of American and European universities, and it is altogether fitting and proper to celebrate so extraordinary a life.

Those were the words I used to honour Peter Medawar's memory at a meeting of his friends in New York soon after he died. I repeat them now to introduce this selection of his writings left behind, made by his friend David Pyke. It is a surprise and pleasure that there are so many. *The Threat and the Glory* will be a delight for those who know Sir Peter's writings and for those who still have that treat to come.

Acknowledgements

I should like to thank the original publishers of these essays for their permission to include them in this volume. The following list gives details of their relevant prior publication.

D.A.P.

1. My Life in Science: transcript of a discussion on the BBC Third Programme broadcast on 25 April 1966 (not previously published).
2. The Threat and the Glory: review of June Goodfield, *Playing God: Genetic Engineering and the Manipulation of Life* (Random House), Michael Rogers, *Biohazard* (Knopf), and Nicholas Wade, *The Ultimate Experience: Man-Made Evolution* (Walker), *New York Review of Books*, 27 October 1977 (under the title 'Fear of DNA').
3. Biology and Man's Estimation of Himself: talk given at a symposium organized by Dr Otto Westphal of the Max-Planck-Institut für Immunobiologie in Freiburg, May 1983.
4. Some Reflections on Science and Civilization: with J. S. Medawar; a talk given at a Ciba Foundation Symposium, published in *Civilization and Science, in Conflict or Collaboration?* (Elsevier/Excerpta Medica/North-Holland, 1972).
5. Florey Story: review of Gwyn Macfarlane, *Howard Florey: The Making of a Scientist* (Oxford University Press, 1979), *London Review of Books*, 20 December 1979.
6. The 'Ultra-Élite' of Science: review of Harriet Zuckerman, *Scientific Elite: Nobel Laureates in the United States* (Collier, 1977), *Minerva*, Volume XV, Spring 1977, pp. 105–14.
7. Scientific Fraud: review of W. Broad and N. Wade, *Betrayers of Truth: Fraud and Deceit in the Halls of Science* (Century, 1983), *London Review of Books*, 17–30 November 1983.
8. The Strange Case of the Spotted Mice: review of Joseph Hixson, *The Patchwork Mouse* (Anchor Press, 1976), *New York Review of Books*, 15 April 1976.
9. Creativity—Especially in Science: written in 1985, unpublished.
10. The Philosophy of Karl Popper: a Richard Bradford Trust lecture given at the Royal Institution on 4 March 1977, published in *Art, Science and Human Progress*, John Murray, 1983.

11. The Genetic Improvement of Man: *The Hope of Progress* (Methuen, 1972).
12. The Future of Man: BBC Reith Lectures 1959 (Methuen, 1960).
13. Osler's Razor: review of Lewis Thomas, *The Youngest Science* (Viking, 1983), *London Review of Books*, 17 February–2 March 1983.
14. The Meaning of Silence: review of Lewis Thomas, *Late Night Thoughts on Listening to Mahler's Ninth Symphony* (Viking, 1984), *London Review of Books*, 2–15 February 1984.
15. The Cost-Benefit Analysis of Pure Research: editorial in *Hospital Practice*, September 1973.
16. The Pure Science: *New York Times*, 24 June 1973.
17. Is the Scientific Paper a Fraud?: unscripted broadcast on BBC Third Programme, *Listener* 70, 12 September 1963.
18. The Pissing Evile: review of Michael Bliss, *The Discovery of Insulin* (Paul Harris, 1983), *London Review of Books*, 1–21 December 1983.
19. Animal Experimentation in a Medical Research Institute: *The Hope of Progress* (Methuen, 1972).
20. Great Circle of Learning: review of the 15th edition of the *Encyclopaedia Britannica*, *New Statesman*, 12 July 1974.
21. In Defence of Doctors: review of Thomas McKeown, *The Role of Medicine* (Princeton, 1980), *New York Review of Books*, 15 May 1980.
22. Son of Stroke: *World Medicine*, 18 October 1972.
23. The Life Instinct and Dignity in Dying: written in 1983, unpublished.

Contents

Introduction

David Pyke

Sir Peter Medawar was three great men. He was a great scientist, a man of great courage—and a great writer. He was supremely creative both as a scientist and as a writer, defining creativity as 'the faculty of mind or spirit that empowers us to bring into existence, ostensibly out of nothing, something of beauty, order or significance'. His creativity in literature was shown in his volumes of essays, especially in *The Art of the Soluble* and *The Hope of Progress*. They consisted largely of reviews, talks, or lectures and, though they were about science and scientists, were written for a general audience.

If I seem to labour the point by saying that he was as great a writer as a scientist it is partly because I agree with him that 'a man's style of writing is an important part of his character—some would say one of the most revealing parts . His recipe for good writing was this: 'Brevity, cogency and clarity are the principal virtues and the greatest of these is clarity.'

Peter was born in Brazil in 1915 of a Lebanese father and English mother. He was sent to school in England and lived there for the rest of his life. When he was still at preparatory school he realized that he was 'hooked on science; no other kind of life would do'. He went to Marlborough, then Magdalen College, Oxford. He got a first class degree in Zoology and then became a Research Fellow.

After exploring various lines of research he focused on the problem of why skin is rejected when grafted from one person to another. He showed that the rejection of skin, kidney, or any other organ is under immunological control. Previously rejection had been considered genetic in origin and therefore insurmountable. But after five years' work he demonstrated, in a series of brilliant experiments, that the barrier could be overcome. The importance of the discovery, which depends upon grafting 'foreign' cells into an animal while it is still *in utero*, was immediately appreciated and led to the award of the Nobel Prize in 1960. It gave great encouragement to the whole medical scientific community and created the new speciality of transplant surgery. Immunologists from all over the world came to work with him and any budding transplant surgeon hoped to have the chance to do so. Peter's discovery of 'immunological tolerance' was important not because it showed *how* rejection by one person of tissue from another could be overcome but *that* it could be overcome. Today's techniques of immune suppression use complex drugs; they would not have been discovered—or not discovered as soon as they were—without Peter's work. He showed the way.

Peter had written about science and scientists for a general audience in *The Uniqueness of the Individual* in 1957 and in the BBC Reith Lectures *The Future of Man* in 1959. After the Nobel Price in 1960 he was asked to speak, review, and broadcast all over the world. He had a great drive to convey to others the meaning and importance of science and to explain his own passion for it. Unlike most scientists he was interested in the philosophy of science, the process of scientific discovery. Science progresses by imaginative leaps which are put to the test in the laboratory: if they fail the test the ideas are discarded, if they pass they survive until later experiments refute or modify them. But what produces the ideas in the first place? They do not enter a vacant mind nor emerge from a heap of randomly assembled observations. 'The most interesting and exciting of all intellectual problems is *how* the imagination is harnessed for the performance of scientific work, so that the steam, instead of blowing off in picturesque clouds and rattling the lid of the kettle is now made to turn a wheel.'

Peter was consumed by science, 'incomparably the most successful activity human beings have ever engaged upon'. It gave him no rest. Scientists are often thought of as problem-solvers. But they are more than that, they are problem-seeking creatures too. 'The motive force that is behind the scientist and technologist's almost compulsive desire for an understanding and mastery of nature is sometimes described as "curiosity" or "inquisitiveness", but these nursery words hardly do justice to what feels like a deep-seated biological impulsion—the hunting feeling, I call it myself.'

Some non-scientists, and perhaps scientists, look with alarm at the progress of science, at the enormous accumulation of facts. They fear that we shall all be overwhelmed by the sheer size of the store of information. On the contrary 'the ballast of factual information, so far from being just about to sink us, is growing daily less . . . In all sciences we are being progressively relieved of the burden of singular instances, the tyranny of the particular. We need no longer record the fall of every apple.'

Peter's writings became famous. One of the most famous was his devastating review of Teilhard de Chardin's *The Phenomenon of Man*. Teilhard was a French priest who claimed to have proved the existence of God scientifically and was widely acclaimed for having done so. But his style was, to put it mildly, opaque. Peter hated opacity in thought or writing. 'People who write obscurely are either unskilled in writing or up to some mischief.' Peter wrote of Teilhard's 'tipsy, euphoristic prose-poetry' in which 'a feeble argument was abominably expressed'. 'It is written in an all but totally unintelligible style, and this is construed as *prima facie* evidence of profundity.' He saw similar pretensions and obscurity in some of the writings of psychoanalysts, whom he condemned for 'the Olympian glibness of their thought, their complete lack of hesitancy and bewilderment in the face of enormously difficult problems. On the contrary a lava-flow of *ad hoc* explanations pours over and around all difficulties, leaving only a few smoothly rounded prominences to mark where they might have lain.'

Much of Peter's writing was light-hearted, but all was serious. Nowhere is this better shown than in the last lines of his review of J. D. Watson's book *The Double Helix*. Watson was the co-discoverer of the structure of DNA, the most important biological discovery of the twentieth century, which explains how characteristics are passed from one generation to the next. His book is amusing, irreverent, naïve, revealing, and vulgar. 'The characters in the discovery of DNA come out larger than life, perhaps, and as different one from another as Caterpillar and Mad Hatter. Watson's childlike vision makes them seem like the creatures of a Wonderland, all at a strange contentious noisy tea party which made room for him because for people like him, at this particular kind of party, there is always room.' Those last fourteen words contain Peter's real message.

Watson was accused of excessive ambition and hunger for priority. Peter saw it differently. A scientist's discovery, unlike an artist's creation, is *his* only in the sense that he made it first. Peter described the difference between artistic and scientific priority in a letter declining to take part in a BBC radio game of scientists versus artists. 'Darwin's claim could not be defended against Beethoven's, even in fun . . . I don't think any scientist can be defended against a major artist; scientists are always dispensable, for, in the long run, others will do what they have been unable to do themselves.'

In 1962 Peter became Director of the National Institute for Medical Research at Mill Hill in London. He set himself to master the job and get to know all the 200 staff at the Institute. He administered the business in three and a half days a week and kept Tuesdays and Thursdays for his own research. Under his direction, the National Institute became a world leader in immunology.

It was as certain as anything can be that he would have been elected President of the Royal Society. This was perhaps the one honour he ever coveted—head of his profession and heir of Isaac Newton. He was riding the crest of a very large wave. There seemed no reason why it should not go on and on.

The wave broke on Sunday, 6 September 1969. Peter, as president, was reading the lesson in Exeter Cathedral at the annual service of the British Association for the Advancement of Science when his voice started to falter. He was helped back to his seat and there sank into unconsciousness. He had had a large cerebral haemorrhage. He was in hospital for months and was near to death twice. While still very ill and when no one knew how badly his mind had been affected he joked about his reading of the lesson in Exeter Cathedral: 'People simply don't know the risks they run when they meddle with the supernatural.' He knew how ill he had been: his reaction was characteristic: 'I myself, naturally sanguine, had considered and dismissed the possibility of dying.'

His experiences in hospital and his close encounters with death did not cause him to bother about the 'dignity' of dying. As he wrote years later: 'No thought of dignity entered my head—it is a state of mind not easily compatible with the hospital microcosm of bedpans and catheters. I needed all the help I could get to promote my ambition to remain alive. It was as allies, then, that I regarded my physicians and the apparatus of intensive care and not as so many plots to deprive me of my dignity.'*

His sanguine attitude to life put him out of sympathy with people such as Ivan Illich and Thomas McKeown† who ascribe all improvements in health over the last century to social and economic factors and dismiss the advances of clinical medicine. 'So long as human beings retain their strong preference for being alive as opposed to being dead so long will medical treatment, if necessary of a strenuous and heroic character, remain in demand. I myself should rather not have needed treatment, but as I did need it, thank God I got it.' Peter got the treatment all right but for the rest of his life he was paralysed in the left arm and leg and could not see anything on the left side of his visual field.

He went back to work in 1970, although not with quite the same ferocious intensity as before. He could not use his left hand, so he could no longer do animal experiments. He started to travel again, always with his wife Jean who, as he put it, was his third leg and arm and eye. His enormous scientific output fell after 1969, but his literary output did not. He wrote seven books (not counting *The Hope of Progress*, published in 1972 but consisting of essays written before 1969). He also gave lectures and radio talks and wrote many reviews.

In 1980, while in New York, Peter had another stroke in the midbrain which affected speech, swallowing, and walking. Again there was gradual recovery but it was not complete. He went on working, writing, speaking, and travelling; supported by Jean, making little of his difficulties.

What a figure he was—6'4", still extremely handsome, half blind, walking with a stick and caliper, his left arm in a sling, his speech a little indistinct but punctuated by laughter. His physical problems were ignored. He was 'heroic in his indifference to increasing disability'. Nothing seemed to impair his enjoyment of life. His

---

* See Chapter 23.   † See Chapter 21.

enjoyment was infectious. He loved to have company, and he was far more sociable than he had been before 1969 when he could seem austere and formidable. His friends loved to be with him. There was none of the reluctance or embarrassment one sometimes feels on being with an old friend who is so afflicted; he accepted the situation as it was and and as he knew it would remain—until it got worse. It did get worse, in June 1985, a few days after he had finished his autobiographical *Memoir of a Thinking Radish*. He had several more strokes until at last, on 2 October 1987, at the age of seventy-two, he died.

When Peter had his third stroke in 1985 there were still many articles and reviews which had not been published in book form. Oxford University Press asked him that year to make a selection for a new book, but by this time he could no longer manage it. It has been an honour to be asked to prepare this book for publication and a labour of love to write this introduction. In doing so, I can say, as Peter did in the introduction to *The Hope of Progress*, that Jean Medawar's help and encouragement have been decisive.

Many of the essays reprinted here have not previously been published in book form. Most are book reviews; I have chosen them from papers Peter left at his death, because they are of general interest and because they do not duplicate others in this or any of the previous volumes.* All but five were written after his first stroke, eight after his second, when he was half paralysed and half blind. Two exceptions are broadcasts, 'My Life in Science' (Chapter 1), not previously published, and 'Is the Scientific Paper a Fraud?' (Chapter 15), a famous talk which appeared in *The Listener* in 1963. The other three have been published in previous books but are now out of print. 'The Genetic Improvement of Man', a paper in honour of his co-Nobel Laureate Macfarlane Burnet, was to have been given by Peter in Australia in September 1969 but had to be read for him; 'Animal Experimentation in a Medical Research Institute', written from his experience at the National Institute for Medical Research at

---

* As the essays were not originally intended for publication in a single volume, there are inevitably a few repetitions and overlaps.

Sir Peter's original footnotes are shown by numbers; the notes that I have added, giving sources or bringing the reader up to date, are marked by symbols.

Mill Hill, is at least as topical and important today as when it was written. Both of these were published in *The Hope of Progress* (Methuen, 1972). 'The Future of Man' comprises the 1959 BBC Reith Lectures. As Peter himself agreed they are tightly argued and therefore solid going, but they are written with his characteristic intensity of purpose, clarity of style, and lightness of touch. After thirty years they are still well worth reading.

I had the great good fortune to know Peter well. We first met in 1959 when his wife Jean joined me in editing a journal on family planning. After his stroke in 1969 when he spent more time at home I came to know him better and to admire him even more. I was once asked if I thought his stroke had affected his intelligence. I said I thought it had: it had reduced his IQ to three figures.

I want to end with Peter's description of the qualities needed by a scientist: 'A sanguine temperament that expects to be able to solve a problem; power of application and that kind of fortitude that keeps them erect in the face of much that might otherwise cast them down; and above all, persistence, a refusal bordering upon obstinacy to give up and admit defeat.'

Peter Medawar wrote that about scientists. It could have been written about himself.

Royal College of Physicians
London

I

My Life in Science
(1966)

WILSON:* Dr Medawar, you were born of a father of Arab extraction, yet went to Marlborough College and then on to one of our older universities. How did this come about?

MEDAWAR: It does sound very romantic doesn't it? But in reality it isn't very mysterious or unusual. My father was a merchant who happened to make his living in Brazil, as many people did. My father wasn't British, he was born in that part of the Arab world which is now called Lebanon, and the Lebanese regard themselves as the heirs of the Phoenicians, and the Phoenicians were great travellers; and he travelled—he travelled to England in fact.

WILSON: And that is how you came to Marlborough College?

MEDAWAR: No. My mother is English and my father was a naturalized Englishman, so it seemed very natural that I should have an English education. My grandfather had been to Marlborough, so it seemed very natural that I should go there too.

WILSON: And was that where you started being a scientist?

MEDAWAR: I started being a scientist at Marlborough. I never seriously contemplated being anything but a scientist, because I went to a prep school in which I had already become extremely interested in science—mainly through reading Benn's Six Minute Booklets. I was very much fired by these books, so I had already

* Sir Peter was questioned by David Wilson, BBC science correspondent, and Dr J. W. N. Watkins, of the department of logic and scientific method at the London School of Economics. I have abbreviated some of the questions and removed a few conversational phrases from the replies.

determined to become a scientist when I went to Marlborough, and Marlborough confirmed it.

WILSON: I suppose the peak of your scientific career was getting the Nobel Prize for certain immunological work. Now can you outline that to us?

MEDAWAR: My mind always goes blank when I'm asked why did I get a Nobel Prize, or how did I get it? Sometimes there's an air of incredulity about it. Ostensibly, I was awarded it with Macfarlane Burnet for a particular scientific discovery; but in reality I think I was awarded the Nobel Prize for having studied the biological theory—the transplantation of tissues—and put it upon a sound experimental footing. The particular discovery was finding out that it is, in principle, possible to overcome the barrier that normally prevents the transplantation of tissues between different individuals. This barrier is an immunological barrier, and at the time we started it was very far from obvious that this barrier could be broken down or that it would ever be possible to transplant tissues from one human being to another. We showed then that it is possible—and in fact, it is now possible by all kinds of different methods, not merely the one that we originally devised.

WATKINS: Dr Medawar, has your scientific thinking, would you say, evolved as it were, gradually and smoothly and continuously, or have there been certain decisive turning-points, any grand climacterics in it?

MEDAWAR: You ask if my scientific thinking has *evolved*. I don't think that is really the right word to use. Given that one is going to be a scientist, what actually turns one's thoughts in one direction rather than another is to quite a large extent luck and opportunity, and the influence of the people you happen to be working with. So I don't detect any natural course of evolution in my work. The immunological work I have just been talking about arose almost accidentally, because at a very early stage of the War I was invited by the Medical Research Council to look into the question of why it was that skin taken from one human being would not form a permanent graft on the body of another. Obviously, because of the gravity of war injuries, it was a matter of particular importance that very extensive skin wounds and burns should be healed by the transplantation of skin. But one could not use skin permanently from voluntary donors and the problem was 'Why not?' I was invited to study this problem, in which I was already interested, and I went into the biological basis of individuality and transplantation.

WATKINS: I did not really mean 'Has your scientific thinking developed along any predetermined course?', but have there been occasions when you leapt out of your bath shouting 'Eureka!', or have things proceeded in a more gradual and piecemeal manner?

MEDAWAR: In a decent, orderly fashion. It could have happened with some people; it does happen. It did not happen to me. That is to say, there was no moment of grand revelation—though there have been a lot of little moments—minor revelations which I think is how scientific work actually proceeds.

The only moment I have ever had in my life where it did appear to me that a new world was opening itself to me was during my undergraduate career. A friend of mine and I went to work in the laboratories of Oundle school to do some scientific messing about of no importance, and I happened to be browsing in the Oundle library, and I came across on the top shelf (I can remember standing on a ladder to get at it) Bertrand Russell's *Principles of Mathematics*. And I read the opening paragraph, which I can still quote verbatim, and I was for some inexplicable reason tremendously impressed by it. I had no idea that anybody could write in that kind of way about mathematics. And that began a lifelong interest in philosophy and logic and mathematical logic. This is the only moment of sudden revelation I have had. Accidental, you see.

WATKINS: Has your interest in philosophy in any distinctive way influenced your scientific work?

MEDAWAR: Yes, I think it has influenced it, and in two totally different ways. I think I picked up from the study of *Principia Mathematica*—I read the first and most of the second volume—a certain—it sounds rather precious to say so—a certain fastidiousness of thought and a certain insistence upon standards of clarity and logical precision and correspondingly a revulsion from any kind of thought which is illogical or irrational. More recently, being interested in the nature of scientific methods, the nature of the scientific process—what goes on in the head when scientific discoveries are made—I obtained, mainly from the writings of Karl Popper, an insight into the nature of scientific discovery which I think has been helpful and could probably be helpful to others too. Now these link two entirely different things. On the one hand the logical and critical side and on the other hand the imaginative and inventive side.

WATKINS: All the philosophical heroes you have mentioned have been modern ones. Have you any older philosophical heroes?

MEDAWAR: I do not think one has philosophical heroes. It is rather dangerous if one does, because one reads into their writings a significance which is not really there. But I immensely enjoy reading older philosophers. At the moment I am reading the works of Dugald Stewart and Thomas Reid, and I delight in their determination to make themselves clear.

WILSON: That is a phrase that I have come across two or three times in your writings: you like hard thinking rather than soft thinking, and you have a great dislike of poetistic prose. Can you define 'hard' thinking as opposed to 'soft' thinking?

MEDAWAR: That is an extremely nasty question. Hard thinking is thinking about particulars or thinking in terms or language that can convey a clear and precise meaning to other people; putting forward ideas which can be tested, which can be the subject of critical examination; statements that make an intellectual appeal as opposed to a visceral appeal—if you admit the distinction. Soft thinking is thinking that makes an appeal to or through the emotions; which gives one a nice cosy feeling inside; which attempts to persuade one of what ought to be intellectual truths by non-intellectual methods.

WATKINS: I wonder whether I'm right in suspecting that it is not merely hot air that you are hostile to, but that sometimes you can be put off what could be regarded as quite a decent scientific hypothesis simply by the fact that it does not satisfy some sort of scientific instinct in you. I get the impression, for instance, that your hostility to Lamarckism is not simply a straightforward scientific objection that the evidence by and large tells against it. I have the feeling that you somehow feel that Lamarckism is a rather tender-minded doctrine, whereas Darwinism is a tough-minded doctrine, and that it is not just empirical facts that govern your choice of theories, but that there is some emotional or aesthetic preference in you. Do you think that is unfair?

MEDAWAR: I think it is unfair. On the other hand I may be deceiving myself about the nature of my own opinions and why I hold some and reject others. I think I reject Lamarckism because it is a view that has not stood up to criticism. I believe it also to be fundamentally mistaken. The genetic system simply does not work in that way. If I have spoken rather vehemently against it it is because it has a certain deceptive appeal. There are some forms of inheritance that are Lamarckian in style, for example the propagation of traditions from one generation to another, a father teaching his children; or learning, learning anything by communication or information from one generation to the next. That is all in the Lamarckian style, and it is very natural to assume that ordinary genetic inheritance is of this kind. But it is not.

To go back to 'hot air', I am quite a believer in hot air in its proper place. I believe that most people psychologically need to be what Paul Jennings calls 'bunkrapt'. (You may remember Paul Jennings's typewriter when he was trying to write 'bankrupt' wrote 'bunkrapt'.) Everybody needs to be bunkrapt, and I prefer to be bunkrapt by listening to Wagner's music dramas or reading Tolkien's novels. It must not spill over into science. It simply won't do. And this is one of the reasons why I have been so indignant about poor Teilhard de Chardin.

WATKINS: Arising out of your paper 'Is the Scientific Paper a Fraud?* which was written under the influence of Karl Popper's ideas on scientific methods your answer was 'Yes, it *is* a fraud' in the sense that it systematically conceals or distorts the way in which the ideas were thought out and developed. Have any of your scientific papers been in this sense fraudulent?

* See Chapter 15.

MEDAWAR: A good many of my papers have been moderately fraudulent. Let me put it this way: I have never written a paper in the complete Popperian style. On the other hand I have never pretended that the research I reported in the scientific paper was done in the inductive style—that is to say by the vacuous collection of facts which then tumbled somehow or other into place. I think I have adopted a compromise. I have not practised what I have preached, but then I am not the first person to fail to do so.

WILSON: That reminds me of the scientific paper which began 'As I was sitting on a rock by the seaside . . .'. Would you go as far as that?

MEDAWAR: Popper himself would never for a moment have condoned writing of that kind in a scientific paper. All that Popper says (I think it is the essence of his opinion) is that scientific activity of any particular kind is preceded by framing in your mind some conception of what might be the case. That is all an hypothesis, an opinion. One invents a possible world and then takes steps to find out whether the possible world one has imagined does correspond, to a first approximation anyway, to the real world. This, of course, does not include the description of what he had for breakfast or whether he was sitting on a rock.

WILSON: The difficulty here is, where do the new ideas come from? You have had new ideas. Where have they come from?

MEDAWAR: The paper you have both been referring to, 'Is the Scientific Paper a Fraud?', was actually an unscripted BBC broadcast, which I gave from fairly full notes, and which was subsequently, much to my surprise, taken up and very widely commented. on. Now one of the things I say in that paper is that there is no logically conclusive process by which you can proceed from particular statements to general statements which contain more information than the sum of their known instances. Now that statement is true.

WILSON: But when you make a new statement, or a conjecture in your own mind, there is something extra to what has been said all around you. That comes from somewhere: where?

MEDAWAR: This is the creative or inspirational act, the nature of which is unknown. All one knows about it is that whatever precedes the entry of an idea into the mind is not known consciously. It is something subconscious. There is a piecing together and putting together of something in the mind but unfortunately of this process nothing is known. There have been attempts to analyse the process. Coleridge, for example, believed that the creative process was a sort of microcosmic reproduction of the original Divine Creation, that out of a formless chaos of words or vaguely formulated ideas poetic ideas arouse spontaneously in his mind. Now it happens that Coleridge has been the subject of a very great deal of critical and exegetic attention, and it is more certainly true of Coleridge than of any other man that his conception of the creative process was not correct, because almost all his ideas can be traced back to something he read in the past, ideas known to have entered his mind. He put them together in a distinctive way but they were not original in the sense in which he used the word 'original'—arising, as it were, out of chaos, out of nothing.

WILSON: Then it is not just a rearrangement of subassemblies into a new order?

MEDAWAR: It is something mechanical in the sense that it is something that could be reproduced by a computer. But no computer has reproduced it and it is very difficult to imagine a computer which would reproduce it economically. One can imagine a computer which could be fed with an enormous amount of information and then, as it were, programmed to devise hypotheses, to account for the facts that had been fed into it. As there are an infinitude of hypotheses which can account for any finite collection of facts, you would also have to impose some restrictions on the computer so that it produced what you might call sensible ideas. We do not always produce sensible ideas, but on the whole scientists have undergone a discipline, perhaps an over-discipline, which ensures that their computers are working pretty efficiently, and they tend to have reasonably sensible ideas to account for the phenomena that they are trying to describe.

WATKINS: Does it not happen in scientific discovery that one breaks through to a discovery of an altogether novel sort of concept, so that the computer analogy of a random process breaks down?

MEDAWAR: I do not think completely new ideas ever arise in that way without some kind of gestation period which one is vaguely aware of. If one has a completely new idea—particularly an idea which overthrows preceding ideas—it is preceded by a certain uneasiness or restlessness which one is aware of. It is a generalized sense of dissatisfaction with all preceding explanations of the phenomena, and eventually you may hit upon an altogether new explanation but then the groundwork for it has been laid.

WATKINS: If one takes a long view surely conservation principles do not work in the scientific sphere? New scientific ideas surely have come into existence; could a computer produce new ideas?

MEDAWAR: This may be a rather sterile subject of discussion, the computer. But I have nothing *against* brains and nothing in particular *for* computers. All I am saying when I say that a computer could conceivably do what a brain does is that there is some rational explanation. It will be interpreted in the language of physiology and neurology in due course. If brains can produce something totally new the computers can produce something totally new too. They can produce an entirely new conjunction of words and ideas, and they may or may not make sense, sometimes they might make sense, if only by accident. That would be something new.

WILSON: As the Director of the National Institute for Medical Research you are faced with the old problem of whether you are going to be a scientist or an administrator. Whichever of those you decide to be you are going to keep up the standard of creativity of your staff there?

MEDAWAR: I construe my function as a director as mainly to create the kind of environment which is conducive to the advancement of learning. That sounds pompous, but this is all a director can do. You cannot *direct* people to have ideas, and no one man can have a big enough grasp of the whole of biological science to be able to say which lines of research are certainly going to be fruitful and which are certainly going to be a waste of time. So what one has to do is simply to create an environment and an atmosphere in which science flourishes. That I take to be my main directoral function.

WILSON: But you pursue your own research, perhaps more than some other directors. Now is this because you feel that this helps you to create the creative atmosphere?

MEDAWAR: Yes. It probably helps to create the right kind of atmosphere if people see that the director, who presumably doesn't *have* to do any research, actually does it and does it because he likes it. I am not saying that this is the best way to run an institute. I don't know the best way to run an institute, I am certain that there are at least half-a-dozen different ways of doing so, others of which might be equally effective—perhaps more effective. But it is the one that suits me.

WATKINS: You have written a good deal about ageing and senescence, and one of the things which you mention is that the faculties can come to their optimum level at different times. To put it bluntly: are you at fifty still in your scientific prime or are you past it?

MEDAWAR: That is a very difficult question for anybody to answer of himself, but I will make an attempt nevertheless. I am quite certainly less quick-witted than I used to be. I am quite certainly less analytically deft. If science is essentially an analytical procedure, if it depended upon let us say, the working of an inductive machine, then I should have to admit that I was going rapidly downhill. I do not wish to deny that for one moment. But another factor has to be taken into account: an important element in scientific discovery is the imaginative element and this does not appear to decay at the same rate as other faculties of the mind. The critical faculty by which one puts one's imaginings to the test may deteriorate, and perhaps it has done in my case, who knows?, but the imagination itself seems to persist. So considering that having good, useful, fruitful scientific ideas is the most precious thing a scientist can have, then I should be reasonably optimistic about continuing to be a scientist. As for the buildup of experience that goes with growing older, this is a very real factor. I have learnt to use my time. I do not believe I waste any time at all during a very long working week. I think younger people do not know how to occupy their time completely, and perhaps when they are young they ought not to be under pressure to do so. I can organize myself and my work very effectively, so perhaps I make good in that way for what must be a certain real loss of capability associated with growing old.

WATKINS: Do you think another advantage is perhaps that you develop a better nose for what will be a promising line of enquiry?

MEDAWAR: I would like to think so but I am very suspicious of that. I think that when people reach an age at which they think they know exactly the right lines for other people to work on they are at a pretty dangerous age. I suspect them in much the same way as I suspect people who pride themselves on being good at summing up other people's characters—they probably do not know anything about character at all.

WILSON: How do you organize your time?

MEDAWAR: I devote some time wholly to administration and some time wholly to scientific research. The unit of time here is days, and I think it can be done. But there is one necessary condition which is sometimes overlooked: one has to *work*—and by 'work' I don't mean working long hours. A great many people think they work hard who merely work long hours. It is a question of working under considerable pressure, working really fast, covering ground and deliberately planning what one does so that one does not waste time. Given that, and given a determination to do research, then it is possible to do research. It has been said that becoming a university professor in England is almost a death sentence from the point of view of doing research, and it certainly is true that a professor has enough administrative work to relieve him of most of the moral obligation if he doesn't actually want to do any research. But if he does want to he can, and nearly all the professors I know do a great deal of research and do it very fruitfully.

WATKINS: In your research, do you actually do physical surgery yourself?

MEDAWAR: I do experimentation myself. I wouldn't regard myself as a scientist unless I actually sat at a bench and did scientific research. The kind of pleasure I get is partly an artisan's pleasure. I like using my hands and I positively and actively enjoy doing scientific research in practice.

WILSON: Some time ago you led a group of people who wrote a letter to the Pope about birth control. Do you believe the responsibility of a scientist stretches beyond the search for pure knowledge?

MEDAWAR: The scientist certainly has a social responsibility and it is a special responsibility where he has, or thinks he has, special knowledge which he thinks ought to influence others in possession of authority. The letter to the Pope I would describe as a rather special case. About eighty or ninety Nobel prize-winners throughout the entire world did sign an exhortation to the Pope, if it is not very impious to attempt to exhort the Pope to do anything. We did so calling attention to the gravity of the population problem and urging that His Holiness should do something about it.

The older image of the scientist was of a person withdrawn from a public life, indifferent to political life. This has completely disappeared now and scientists have gone perhaps to the other extreme. I do not think any scientist can or ought to repudiate his unit responsibility in a democracy but I must repudiate the idea that the scientist commands a method, the scientific method, which gives him a privileged and special insight into social affairs or into the solution of social problems. This was Mill's belief; John Stuart Mill wanted to find out what the scientific method was in order that he should apply this wonderfully successful method to solving the problems of society. Now Mill's method failed completely. In fact there is no scientific method which could possibly be applied to the resolution, as a whole, of social problems. Scientists are sometimes inclined to feel that if only they could have positions of great political authority and exercise the scientific method then all problems would dissolve before their acute critical gaze. It does not work like that, it just is not true.

WATKINS: Have you ever felt at all that a certain line of enquiry ought to be stifled because it might lead to very dangerous results?

MEDAWAR: That is a very difficult question to answer in abstract terms because if a scientist takes that kind of decision it means that he thinks that the only evidence relevant to making this decision is scientific evidence. Now this is almost always false. There are very few modern problems to which scientific evidence is irrelevant, but there can be very few problems where scientific evidence is all the evidence you need. The example that is always taken is should the early workers on the atomic bomb have taken it upon themselves not to develop the bomb? That decision would have been an extremely grave and far-reaching political decision which might have been disastrous to many of the things which the atomic scientists themselves held to be sacred. It is not a scientific decision. On the other hand, the scientific evidence is immensely relevant to that decision. The scientists might well have said—and did say—what an immensely destructive weapon this was. But to go beyond that and take political decisions seems to me to be entirely wrong. This is not to say that politicians make the right decisions, but they are more likely to make right decisions than scientists simply because they know more of the other factors that are relevant to making a decision.

WATKINS: There are people who, considering the evolutionary time-scale and reflecting on the fleeting existence of each individual organism (it comes into existence, passes on its genetic code, and is then extinguished), may be frightened by this vast and apparently impersonal process. Do you share that feeling at all?

MEDAWAR: No. I think what actually happens in nature—the passing on of the genetic code—the continuity, but at the same time the intermingling of the genetic factors that play such a large part in controlling and determining personality, is a grander conception than one which is founded wholly upon the preservation of the individual.

WATKINS: Isn't it a little frustrating to realize that we are purely vehicles for something which is immortal rather than participating in it ourselves?

MEDAWAR: I think we do enjoy it, and we do participate in it. For example, when one sees one's own interests or aptitudes developing in one's children; if one is a teacher, when one finds that something one has taught to other people is growing in their minds, you are making your own personal impact upon the genetic flux that goes through generation after generation. A fragment of your influence therefore does become immortal. If one is conscious of this process, that one is taking part in a huge genetic process, this is a source of comfort and enjoyment and even of wonder and awe.

WILSON: Is that all that you think is the immortal? And if so, is it your work that specifically leads you to that conclusion?

MEDAWAR: Certainly not. No, I think this is just a general biological conception which a great many biologists would probably agree with.

WATKINS: Do you agree that, so far as human beings are concerned, it is not so much genetic evolution that counts as the other sort of evolution you referred to—the transmission of ideas, books, facts, knowledge?

MEDAWAR: Knowledge and know-how, yes. We contribute to this on the one hand and we contribute to the genetic heritage of succeeding generations on the other hand. Of course, this genetic contribution may not be a good one, and the contribution we make through exogenetic agencies, through the medium of tradition, may also be a bad one. In that case whether it's good or bad does depend to a very large extent on ourselves. We can more easily have a good influence through teaching and by example then we can have a good genetic influence, and so we have some control over that moiety of inheritance which is not genetic, or which is exogenetic. This is a great satisfaction and one that people are very dimly aware of. They don't think of it as a fragment of immortality in themselves. They think of it as something different, but in a way it is a kind of immortality.

WILSON: And is this where moral responsibility comes from?

MEDAWAR: I should not go as far as to say that. I have never thought of that question, but my instinct would be to say no.

WATKINS: Has your scientific thinking influenced your moral or political, perhaps even your aesthetic outlook?

MEDAWAR: As to aesthetic outlook, I would say not at all. Morally no, I do not think so either, though that is a good question because the scientists are so very often on what one might call a 'progressive' or 'liberal-minded' side in politics rather than a narrowly conservative one. One would think that there would be some connection between the pursuit of science and having liberal ideas. If there is such a connection, I don't know what it is—unless the scientist is from the nature of his work more imaginative. He is used to solving problems, and he may therefore think that the major problems of society will yield to solution by methods as comparatively simple as those that he uses in his laboratory.

WILSON: Your work is connected with immunology and with the problem of individuality. Our immunology is the definition of our uniqueness as individuals. Is it possible to imagine a time when the success of your own work in immunology will undermine our uniqueness immunologically?

MEDAWAR: So that a time will come when spare parts will be freely interchangeable for everybody, as between model-T Fords?

WILSON: Could you even push this to the point of not starting new human beings; you just keep one going—by continuously transferring fresh organs?

MEDAWAR: The point of the questions is where does personality and sense of individuality reside? You can imagine a human being whose every organ, except perhaps one, came from some other human being; now would he remain himself or would he be somebody else? I don't think physical integrity plays a very important part in one's conception of identity or individuality. A man is diminished in John Donne's sense by losing a leg. He is physically diminished, but his persona is not diminished. It is one's store of memories that I suppose make one different from anybody else, and these are stored somehow or other in the mind. As the seat of the mind is the brain so the brain is clearly the organ we must preserve if we wish to preserve our own identities. The rest I do not think matters very much.

2

The Threat and the Glory
(1977)

It is the great glory as it is also the great threat of science that everything which is in principle possible can be done if the intention to do it is sufficiently resolute. Scientists may exult in the glory, but in the middle of the twentieth century the reaction of ordinary people is more often to cower at the threat.

Everybody will doubtless be dismayed to learn that it is possible in principle—and technically not even very difficult—to transform human beings into two sub-peoples: the one moiety brainy and comparatively beautiful—like the Eloi of H. G. Wells's famous journey into far future time—and the other moiety comparatively stupid but fitted by their docility and physical strength to do the dirty work and serve the others: Wells's Morlocks or Wagner's Nibelungen.

Why does not the mere possibility of this ultimate political prostration of mankind fill us with dismay? The reason is that the programme I have just envisaged could have been embarked upon at any time in the past thousand years, merely by applying the most powerful of all forms of biological engineering—Darwinian selection—to a population—mankind—known by its open breeding system, lack of specialization, and rich resources of inborn diversity to be perfectly well able to respond to the empirical arts of the stockbreeder. The answer, in the form of a counter-question, does something to explain why most biologists and laymen look rather coolly upon such attempts to curdle our blood: if these enormities have not been perpetrated or even seriously attempted hitherto by the comparatively straightforward and empirically well-understood methods available for their execution, why should we now begin to fear that enormities as great or even greater will be executed by the much more costly and technically more difficult procedures of genetic engineering—by procedures which are conceptually well understood, to be sure, but are not yet anywhere near the level of proficiency in actual execution which the stockbreeder can command?

Nothing since the early days of atomic weaponry has caused so much dismay as the real or imagined threats associated with the development of genetical engineering and recombinant DNA research, the subjects of the books under review.*

At the root of all genetical engineering lies that which I described without qualification as the greatest scientific discovery of the twentieth century: that the chemical make-up of the compound deoxyribonucleic acid (DNA)—and in particular the order in which the four different nucleotides out of which it is assembled lie along the backbone of the molecule—encodes genetic information and is the material vehicle of the instructions by which one generation of organisms governs the development of the next. If the DNA message is altered the effects of doing so are, in their context and of their kind, as far-reaching as the effects would be of altering the wording of Congressional or Parliamentary legislation or the wording of telegrams conveying diplomatic exchanges between nations. It is just such a process as this which in recent years has become possible by direct intervention—and to some degree at the experimenter's will—a situation quite different from the action of natural or artificial selection upon naturally occurring differences in the DNA messages characteristic of different organisms. The first process changes the genetic make-up of an organism, the second changes the make-up of a population of organisms.

Introducing what has become the most talked about version of the first process—'recombinant DNA'—June Goodfield comments, 'Very simply, it is the new technology that enables a scientist to take DNA from one organism and splice it onto DNA from another to create something absolutely new: new living molecules, new genes, and therefore new life.'

* June Goodfield, *Playing God: Genetic Engineering and the Manipulation of Life* (Random House, 1977), Michael Rogers, *Biohazard* (Knopf, 1977), and Nicholas Wade, *The Ultimate Experience: Man-Made Evolution* (Walker, 1977).

The term 'biological engineering' need not of course be confined to that part of it which takes the form of an attempted manipulation of DNA. 'Engineering' embraces all that accompanies and makes possible the translation of thought into action, and even if 'thought' is too far-fetched a description of the acts of mind that underlie some of its manifestations, 'biological engineering' can certainly be extended to include suspension of life in the deep-freeze, the attempt to rear babies to term outside the body, and other enterprises upon which the Medawars[1] have not thought 'idiotic' too harsh a judgement.

Francis Bacon described the goal of the New Science of the seventeenth century as 'the effecting of all things possible'. The agents of this tremendous ambition were to be wise men and philosophers; he did not think there would ever come a time when people would do things merely because they *were* possible, yet that is exactly the mischief which the biochemist Erwin Chargaff, whom June Goodfield quotes, describes as the devil's doctrine: *what can be done, must be done.* It must have been some recognition of this source of temptation in themselves or in their weaker brethren that led to the remarkable resolutions of the Asilomar Conference of February 1975 in California at which scientists themselves proposed that certain types of experimentation with DNA should be abstained from. No literary folk have ever done as much. On the contrary: any suggestion that an author should not write exactly as he pleases no matter what offence he causes or what damage he does is greeted by cries of dismay and warnings that any such action would inflict irreparable damage on the human spirit and stifle for ever more the creative afflatus. Let us count it a mercy that we do not have to put up with this kind of talk from scientists, I mean, put up with the argument that the discovery of the truth is a complete justification for anything they may choose to do.

Although it was historically the most important, the Asilomar Conference of 1975 is not the only evidence of an awareness of possible evils acute enough to prompt scientists to accept guidance or impose upon themselves a censorship restricting their freedom to

---

[1] P. B. and J. S. Medawar, *The Life Science* (Harper & Row, 1977).

do exactly what they please. The National Institutes of Health [NIH] have issued guidelines on the prosecution of recombinant DNA research and the British Medical Research Council has issued a cautionary document on genetic manipulation guided largely by the report of Lord Ashby's Working Party on this subject.[2] The Federation of American Scientists [FAS] has issued a thoughtful and gravely worded public interest report[3] on the subject and the New York Academy of Sciences has devoted a symposium volume to a conference on the ethical and scientific problems raised by the human uses of molecular genetics.[4] At this conference Daniel Callaghan asked 'How, then, are we to possess power without being possessed by it?', adding that this was the fundamental question underlining the problem of ethical responsibility in science. Lord Acton and others have pointed out that the same is true of political action. Callaghan is not one to blame the weapon for the crime and he says that 'if the quest for scientific knowledge is to be condemned because some of that knowledge may be misused, then so must the quest for all knowledge'. Again, 'there is no special responsibility applying to scientists that does not apply to others'.

There was this difference though: scientists were now more fully cognizant than ever before of the way in which innocent-seeming and intrinsically inoffensive experimentation may lead to disastrous consequences. It was therefore, Callaghan said, a special obligation upon a scientist to envisage what consequences of his work were *conceivable* and to share these misgivings with his colleagues. I believe that it is just this attitude which underlies the present unease of biologists about what the consequences of molecular genetic engineering may be.

In his book *Biohazard*, Michael Rogers does not plunge us right into the middle of things but explains carefully and intelligibly the classical researches that provided the conceptual foundations of modern genetic engineering, making special mention of Archibald

---

[2] *Report of the Working Party on the Experimental Manipulation of the Genetic Composition of Micro-organisms*, Cmnd. 5880 (London, Her Majesty's Stationery Office, 1975), pp. 11–12.

[3] *FAS Public Interest Report*, 29/4 (Washington, DC, Apr. 1976).

[4] *Ethical and Scientific Issues Posed by Human Uses of Molecular Genetics*, Annals of the New York Academy of Sciences, 265 (Jan. 1976).

Garrod, who first identified the so-called 'inborn errors of metabolism' that occur because the body has a missing or defective gene, and of the classical experimental researches of Beadle and Tatum on the bread mould *Neurospora crassa* showing the connection between the action of genes and that of enzymes. Garrod's work and the *Neurospora* work represent some of the finest science of the twentieth century. From there he proceeds, justly and inevitably, to the dramatic and often recounted work on pneumonia bacteria by O. T. Avery and his colleagues in the Rockefeller Institute. These brilliant experiments first revealed that the gene-like agent responsible for transforming certain bacteria from being non-lethal to lethal was none other than deoxyribonucleic acid—DNA for short—an abbreviation Rogers is sanguine enough to believe has now entered the vernacular. It is especially pleasing to see the prominence given to the name of a man, O. T. Avery, who deserves type as big and lights as bright as those of anyone who helped to tell the great story of DNA. Rogers, Wade, and Goodfield tell the same story of course: it is a good story and all three tell it well and in much the same way, though Goodfield's *aperçus* are the most personal.

It will now be helpful to take evidence from a variety of different well-informed sources.

Nature, the world's foremost scientific newspaper, has not stood aloof from the controversy. On the contrary, looking back over 'Recombinant DNA Debate Three Years On',[5] an editorial declares that:

information generated during the past three years indicates that the potential hazards associated with gene-splicing experiments may be more remote than once believed. For example, a special meeting of scientists and health experts, convened by NIH earlier this month, concluded that there is virtually no chance that recombinant DNA experiments could touch off an uncontrollable epidemic.

Nature goes on to cite Dr Roy Curtiss, a respected microbiologist from the University of Alabama, as having written after much experimentation with the laboratory strains of the bacillus *E. coli* that are being used in genetic research: 'I have gradually come to the

[5] *Nature*, 268 (21 July 1977), 185.

realisation that the introduction of foreign DNA into EK1 and EK2 host-vectors offers no danger whatsoever to any human being.'

A more serious danger, maybe, is that the allegedly hazardous nature of the work may induce grant-giving agencies to impede the development of molecular biology or, more likely, to give molecular biologists seemingly valid reasons why their patrons should pull the purse-strings together just when authentic supplicants are peering eagerly inside. A statesman-like frown is accordingly directed at Senator Edward Kennedy's health subcommittee which is engaged in devising restrictive legislation that could possibly impede worthwhile research.

The Federation of American Scientists has a long record of service to the community, and the article 'Splitting Atoms and Transplanting Genes' in its recent *Public Interest Report* very properly reminds us of its stalwart services to the nation in making sure that the hazards of atomic energy became widely known. It now sees it as part of its function to do as much for recombinant DNA research, but so far from holding up the profession to public obloquy the FAS writes of it rather handsomely:

The researchers have behaved with unprecedented restraint and caution. Raising the issue themselves; bringing it to public attention; urging the voluntary deferral of various experiments; and debating the hazards in full public view, represents four quite different and thoroughly commendable steps. In addition, most have, quite surprisingly, been able to come to agreement on a set of guidelines that have grown steadily more stringent— even while many of the researchers have grown more sanguine about the dangers. This is a tribute to the statesmanship of their leaders. It is no surprise that now they want to go ahead with research which all observers agree is filled with promise, and which promises tremendous assistance in understanding biology. They only ask a 'yellow' light—the right to proceed with caution.

Among the hazards the FAS calls attention to is the accidental escape of potentially dangerous organisms as yet unknown in nature. The FAS seems to fear that the body's immunological system would be confounded by such unknown organisms. This fear of the unknown because it *is* unknown is not really justified. Human beings for example are perfectly capable of mounting immunological reactions against organisms new to them or even against chemical compounds which they have never met before—which, indeed, have not yet been invented. It is a misunderstanding of physiology to suppose that immunological-like neurological reactions depend at least in part on prior experience: there is after all always a first time we are confronted with any disease-causing organism but we do not necessarily succumb to it. The FAS goes on to say:

The basic current hazard is the introduction into bacteria of genes which make the bacteria more dangerous. In the simplest case, such genetic changes might give one strain of bacteria the resistance to antibiotics that exists in other strains; thus some such antibiotic as penicillin might suddenly find that strains of bacteria that cause pneumonia had become resistant to its application.

It is notorious, though, that this process has been going on since penicillin was first introduced into medical practice and used more frequently and in larger doses than immediate needs called for. The appearance of antibiotic-resistant strains of formerly susceptible bacteria is a typical evolutionary process. Although the existence of penicillin-resistant strains of bacteria is a major nuisance, it does not portend widespread disaster: rather it puts biologists on their mettle to find ways around the problem.

The FAS draws special attention to and endorses the main conclusions of the Working Party under Lord Ashby of the benefits and possible risks of genetic engineering. Its conclusions are worth setting out anew:

We now have to declare our assessment of the potential benefits and practical hazards of using the techniques we have described. We reiterate our unanimous view that the potential benefits are likely to be great. The most substantial (though unpredictable) benefit to be expected from the techniques is that they may lead to a rapid advance in our detailed understanding of gene action. This in turn might add substantially to our understanding of immunology, resistance to antibiotics, cancer, and other medically important subjects.

Furthermore, application of the techniques might enable agricultural scientists to extend the climatic range of crops and to equip plants to secure their nitrogen supply from the air. Another possible application is that segments of DNA, selected because they are templates for valuable products such as hormones, antigens or antibodies, might be produced in bulk by multiplying them in cultures of *E. coli*:* this would be of great benefit to medicine. And it is not inconceivable that the technique might ultimately lead to ways to cure some human diseases known to be due to genetic deficiency.

In discussing the hazards of these techniques we have to distinguish between the risk to workers in the laboratory and the risk to the public. Many scientists are engaged on potentially hazardous research (using radioactive materials, or unstable chemicals, or pathogens). They and those who work with them are trained to take precautions; accidents are rare and they do not spread. But if the danger is one which might not be contained within the laboratory, the need for precaution is much greater and the public have a right to seek assurances that they are not at risk.

Fortunately there are precedents for making such assurances. In the production of some vaccines, in public health and hospital laboratories, in research institutes for the study of infectious disease, it is essential to handle pathogenic organisms, some of them extremely dangerous. Accordingly, there is a well developed strategy of containment for these hazardous operations . . . The dramatic response to any failure in containment illustrates how rare such failures are. A recent example of this is the enquiry which followed an outbreak of smallpox in London in 1973.

In short, the potential benefits of recombinant DNA research are great and the precautions the experiments call for must be commensurate with the magnitude of the risks involved.

I once had the pleasure of a lengthy formal discussion with Jacques Monod, at that time the Director of the Institut Pasteur, about a number of biological problems having to do with the threat and promise of genetical engineering—a subject upon which he was as well qualified as anybody in the world to express an authoritative opinion.† We agreed that both the threats and the promises were greatly exaggerated and that the realization of both good and bad dreams was a very much more difficult exercise than it was commonly assumed to be, but at the same time Monod made an exception of cloning—the production of an indefinitely large number of replicas of some chosen human type.

Cloning was a definite possibility, he believed. The procedure has nothing to do with the recombinant DNA, however; it is biological

---

* This is now a reality.   † See the *Listener*, 3 Aug. 1972.

engineering in the wider sense discussed above. To get it into perspective I should like to run over the procedure that would have to be adopted if cloning were to succeed. The first step would be to wash out from the fallopian tube a fertilized and therefore activated human egg—an egg developmentally ready to go. The egg would be stored in a cool, sterile nutrient medium outside the body and then manipulation could begin. If toads and newts are anything to go by, the egg's own nucleus could be replaced by a nucleus from an ordinary body cell (a lymphocyte nucleus, mayhap) from the tissues of the individual chosen for indefinite replication. The egg would then be maintained under conditions which allowed it to undergo a number of successive cell divisions—a process almost exactly analogous to twinning as it may sometimes occur *in vivo*.

That would only be the beginning of it, however; because for each such daughter egg to develop into a human being it would be necessary to find a woman whose uterus had been prepared by hormones in such a way that the daughter egg transplanted into it would continue with cell division and eventually attach itself to the uterine wall—'implantation' is the technical word. The embryo might or might not go to term; if it did, it would necessarily have the same genetic make-up as the individual whose cell nucleus substituted for the nucleus of the original egg.

Anybody with any experience of experimental pathology—and the rival attraction of less exacting pursuits means that their number is getting less and less—knows that to carry through this programme and to overcome all the misadventures that could so easily befall it, would require a degree of organization that would make the mobilization and deployment of an army seem like running a Sunday school picnic. Even supposing a grant-giving agency composed mainly of wealthy simpletons could be found to fund such a foolish enterprise, the very many misadventures known by all experimentalists to beset such a scheme would almost certainly prevent its being realized. We need not worry then about the difficulty of finding any one human being whose indefinite replication could be thought of with equanimity, for considered as a whole the enterprise is simply not on.

No appraisal of genetic engineering would be fair unless Erwin

Chargaff were called to the witness stand. Chargaff was one of those who played a leading part in the discoveries that led to our modern understanding of DNA, and his part too, like Avery's, is not as well known as it ought to be. In writing 'On the Dangers of Genetic Meddling'[6] Chargaff is very sceptical about the overflowing cornucopia of advances in medicine and human welfare which, it has been alleged, grow out of the use of gene-splicing techniques—benefactions thought to include the repair of human genetic defects (a procedure very far beyond our present competence). Of this project I have said:

> It is mentioned in the same spirit as that in which a young biologist seeking funds to study the growth of sea cucumbers in a pleasant seaside resort urges his patrons to believe that such an investigation will throw a flood of light on the transformation from the normal to the malignant cell: it is a harmless form of window dressing that all grant-giving bodies understand and allow for.[7]

Chargaff declares that the genetic engineers are not nearly as proficient as they are given the credit for being about the splicing of eukaryotic DNA into DNA of micro-organisms. (Eukaryotic DNA is the DNA of organisms, such as animals and higher plants, in which genetic material is marshalled and structurally organized in the form of chromosomes, the nucleic acid being combined with a basic protein to form a salt-like compound.) 'Most of the experimental results published so far in this field are actually quite unconvincing . . . it appears that the recombination experiments in which a piece of animal DNA is incorporated into the DNA of a microbial plasmid are being performed without a full appreciation of what is going on.'

Chargaff is sceptical of the long-term efficacy of orthodox containment procedures for the possible escape of pathogens and he asks why molecular geneticists have chosen as the subject of their experiments an organism *Escherichia coli*, the colon bacillus, which has for so many millennia been living in a state of symbiosis with man. 'The answer is that we know so much more about *E. coli* than about

---

[6] *Science*, 192 (4 June 1976), 938–40.
[7] 'The Scientific Conscience', *Hospital Practice* (July 1976), 17.

anything else, including ourselves.' He is right: so much knowledge and know-how is vested in E. *coli* that there is little likelihood of its being supplanted as a subject of experiment. In any event, so the patrons of E. *coli* argue, the laboratory organism has now been so modified in the course of prolonged culture outside the body that it no longer qualifies to be considered a regular member of the flora in our gut.

Clifford Grobstein,[8] well known for his sensible and temperate views, deplores the polarization of the recombinant DNA debate into an antithesis between 'best-case' and 'worst-case scenarios'. The worst-case scenario he envisages comprises: world-wide epidemics caused by newly created pathogens; the triggering of catastrophic ecological imbalances; the power to dominate and control the human spirit.

The last of these imagined dangers rather surprised me, for it seemed to me to be on all fours with H. G. Wells's Eloi/Morlock bad dream referred to earlier. Certainly nothing much more horrible can be envisaged than a procedure which not only fills the mind of man with untruth and misconception but leads to an active resistance to new learning and to anything that might conduce to improvement. Yet here again the technology that puts these grim possibilities within our power has also been known for five thousand years or more: it is known as 'education', and it too has its brighter side, for whatever procedures may persuade us to approve evil can in principle also be used to make us reprobate evil and rejoice in and embrace the good.

Writing of the disquiet of the laity Grobstein makes it clear, though, that 'the fear is not so much of any clear and present danger as it is of imagined future hazards'. Grobstein fears that physical containment and the associated safety precautions reveal something of a Maginot Line mentality, for what is needed is research that will evaluate these hazards precisely, so that we know where we stand and shall not find ourselves standing still.

James D. Watson is well known to have a somewhat messianic

[8] 'The Recombinant-DNA Debate', *Scientific American*, 237/1 (July 1977), 22–33; 'Recombinant DNA Research: Beyond the NIH Guidelines', *Science*, 194 (10 Dec. 1976), 1133–5.

conception of his role in the great revolution of molecular genetics and it was hardly to be expected that he would remain silent amidst the clamour of discussion on recombinant DNA. He says that the Asilomar Conference made him uneasy and he now declares:

I did not then, nor do I now, believe that all recombinant DNA research is necessarily totally safe. The future automatically entails risks and uncertainty, and no sane person rushes in directions where he anticipates harm to himself or others. Instead, we try to adjust our actions to the magnitude of risk. When no measurement is possible because we have never faced a particular situation before, we must not assume the worst. If we did, we would do nothing at all.[9]

I do not think Watson is being unduly sanguine and I specially applaud his choice of the word 'sane'.

Having now taken evidence from various quarters we may turn to the three works specifically under review. Nowadays laymen need not be told that 'Cry havoc!' attracts more attention than the nightwatchman's reassuring 'All's well, all's well.' Happily none of these three books is disfigured by sensationalism; however there is something a little breathy about them all. None of them is definitive or pretends to be: these are interim reports: a definitive treatise could only be written from a height which none of the three authors can command.

Goodfield, though, turns her lack of inside knowledge to advantage by describing how she apprenticed herself to a laboratory in which recombination experiments were taking place. I liked specially her delighted description of the winding out of the exquisitely beautiful DNA fibres on a glass rod after they had been precipitated from solution by the addition of alcohol. It is not an essential part of her narrative, of course, but I sympathize entirely with her wanting to bring it in because when I myself prepared DNA for immunological purposes I can remember cruelly boring my colleagues by calling upon them to witness the very process June Goodfield describes.

Nicholas Wade might say that this episode illustrates his contention that 'Gene splicing is so simple a technique that for most present purposes it requires only a few dollars worth of special materials, all

---

[9] 'In Defense of DNA', *The New Republic* (25 June 1977), 11–14.

commercially available, and access to standard biological laboratory.' I think this is a misjudgement that reminds me of a prominent sociologist's published contention (I shall not say where) that the manufacture of atomic bombs now lies within the capabilities of a high-school student. It could equally well be said that appendectomy is a remarkably simple operation requiring no more facilities than are available in a quite ordinary hospital. But its execution requires a knowledge and know-how—the biological or surgical equivalent of worldly wisdom—which puts it for all practical purposes far beyond the reach of any ordinary villain or casual mischief-maker—a villain who collected appendixes as others collect stamps.

June Goodfield's account has the merit of making it clear by implication why the conferment of antibiotic resistance is such a favourite exercise with genetical engineers. The reason is that it is not much good doing an experiment or modifying its procedure unless one knows whether the experiments work, or work better than before. When the conferment of antibiotic resistance is the transformation attempted, the organisms in which the transformation has been successful can be isolated very easily from a population that may be as diverse as the population of Times Square on a Saturday night (Goodfield's image).

Each of these three books is good and since there is general agreement in the nature of the promises and the threats it would be idle to single out any one of them; for each has special merits. They agree, too, on the history of discoveries bearing on DNA, though Rogers goes back as far as Miescher in the 1870s—the man who first extracted the stuff long called nuclein from pus (one good mark, if we were having a competition). This historical excursion will certainly earn him the contempt of those semi-literates who regard any work done earlier than in the past year or two as of merely antiquarian interest.

Writing of safety precautions in laboratories handling potentially dangerous materials such as tumour viruses Wade quotes W. Emmett Barkley, the biological safety expert at the National Cancer Institute, in these terms: 'In the majority of labs we visit we see things that ought to be corrected. The greatest offenders are university labs, not industrial labs. Most people working with tumor viruses have been exposed to some extent.' Barkley's is the cry of safety officers throughout the world—in factories no less than in laboratories. I offer it gratis to some graduate actuary or sociologist on the lookout for a Ph.D. degree that he should study the life expectancy of safety officers in factories and laboratories. I suspect they die prematurely of diseases of stress.

Incorrigible though their clients seem to be, however, we must keep it firmly in mind that for every steel worker who falls into the blast furnace, and every cider-maker who dissolves boots and all, in raw apple juice (rich in frighteningly powerful enzymes), hundreds and hundreds do not. The parallel is not facetious, because no one is more gravely and immediately at risk of the hazards to which they are believed to be about to expose the public than the people who actually carry out supposedly hazardous experiments. I don't think the general public need take grave alarm until the inmates of institutions of genetic engineering themselves begin to fall by the way.

A further consideration that will influence the worldly-wise is that genetic engineers would very much *like* to confer upon micro-organisms the ability to manufacture, in copious amounts, human insulin and the anti-viral agent called interferon, now being used in the treatment of some cancers.*

When the engineers have demonstrated to everybody's satisfaction that they can do on purpose what they very much want to do, then will be the time to reappraise very critically the dangers consequent upon their inadvertently doing what they do not want to do anyway.

The large-scale manufacture of either human insulin or interferon would be a very great benefaction to mankind, for the trouble with interferon at the moment—so often judged therapeutically disappointing—is that there isn't enough to give it in dosages large enough for a clinical trial of adequate scale. Even penicillin did not finally triumph until it became possible to administer it in doses of the order of mega-units.

In Wade I came across for the first time the idea that nitrifying

---

* They now can.

enzymes might conceivably be incorporated into plants that normally lack them, thus making it possible for them to capture from the atmosphere the nitrogen necessary for their growth and making them independent of added fertilizers (which are essentially compounds of nitrogen). The notion is not impossibly far-fetched because some plants can be raised into whole organisms from single isolated cells. But here too I do very deeply sympathize with laymen and legislators who are trying to make sense of this whole strange farrago of pipe-dreams and nightmares.

For their excess of fearfulness, laymen have only themselves to blame and their nightmares are a judgement upon them for a deep-seated scientific illiteracy which manifests itself in two ways.

In the first place the public deserve nothing but contempt for allowing themselves to be dupes of that form of science fiction which is our modern equivalent of the Gothic romances of Mary Shelley and Mrs Anne Radcliffe; for being taken in, that is to say, by that trusty serio-comic character, the mad scientist who to the accompaniment of peals of maniacal laughter, cries out with a strong Central European accent, 'Soon ze whole vorld vill be in my power.'

The second reason for their excess of fearfulness is this: that because imaginative writing is the only form of creative activity most people know, even educated laymen have no idea of the width of the gap between conception and execution in science. A writer who hits on a good idea—or even a composer who thinks of or, like Sullivan, overhears a good tune—can take up pencil and paper and write it down; he does not have to sue for bench space in a laboratory or send in five copies of an application explaining what his poem is going to be about, how many sheets of paper it will occupy, what imagery it is going to be clothed in, or how mankind will benefit by its completion. But when a scientist has an idea he has merely reached the beginning of a long haul which will certainly involve an appeal for funds which he may easily not get. He cannot simply walk into his laboratory with a purposeful and dedicated look on his face and execute the idea he has in mind.

The existence of this large gap means in effect that the execution of recombinant DNA research depends very largely upon political decisions. I do not use the word 'political' in the sense that it would depend upon Congressional or Parliamentary legislation but simply in the sense that the project and the means of executing it depend on decisions that are not the scientist's alone: they will depend at least in part upon peer judgement and on the policy decisions of an independent grant-giving body. But, it will be objected, many of those responsible for the decisions are themselves scientists; all right, but if one mad scientist is rare, a committee of scientists, all mad, is very much more improbable still. The existence of this very wide gap between conception and execution is that which allows the interposition of wiser counsels and restraining hands between the scientist's idea whether bright or foolish and the possibility of its being put into effect.

So much then for the aetiology and cultural history of the forebodings that cause so much disquiet among laymen. To the professional scientist these suspicions of his competence and probity are most disquieting. In one of a number of wise discourses on civilization Sir Kenneth Clark remarked that all great advances in civilization are based upon *confidence*. Although I have tried to explain it I find it difficult to excuse the lack of confidence that otherwise quite sensible people have in the scientific profession, among whom sanity is much more widely diffused than seems to be generally realized. Scientists want to do good—and very often do. Short of abolishing the profession altogether no legislation can ever effectively be enforced that will seriously impede the scientists' determination to come to a deeper understanding of the material world.

3

Biology and Man's Estimation of Himself
(1983)

Twice in its history biology has seemed to threaten our *amour propre* and to diminish ourselves in our own estimation. I shall describe how this came about and why it was that the threat came to nothing. I shall then briefly discuss the morality of the prolongation of life, by medical engineering or other means.

It has been said that the greatest cultural shock mankind ever experienced arose out of the recognition of anthropoid apes in the first great voyages of discovery in the sixteenth and seventeenth centuries. Chimpanzees (baby chimpanzees were described as 'pygmies') were reported from East Africa and the 'wild man of the woods' that Linnaeus was later to name *Homo sylvestris orang-outan* from South-East Asia. The apparent affront they embody lies in their being caricatures of ourselves—unmistakably similar to ourselves in various ways like an ill-natured political cartoon.

Yet no shock was experienced. If the man-like apes had caused a deep soul searching or revaluation of ourselves, there would have been evidence of it in contemporary literature just as in Ibsen and Shaw there is a frightened awareness of the newer knowledge of heredity. Shakespeare, like Aristotle, probably never saw an ape (Aristotle refers to 'the Barbary ape'—but this is merely a macaque monkey). Shakespeare used 'monkeys' and 'apes' interchangeably. For Shakespeare, apes are nothing worse than mischievous and unduly garrulous. Pascal has no thoughts on them. The only remark of Bacon's known to me is very characteristic of this witty and reverent man: 'As it addeth deformity to an ape to be so like a man so the similitude of superstition to religion makes it the more deformed.'

Why did the apes cause so little stir? The reason, I believe, is that their existence was half expected in the light of a conception that had sunk deeper than we realized into public consciousness: Plato's and Aristotle's great chain or ladder of being. Plato had declared that a rational divinity would surely order the world in such a way as to make an unbroken gradation from the lowest to the highest living thing—from water flea even to divinity. Any creature that could be rationally imagined must exist in reality (the 'principle of plenitude'), and after all it is not at all difficult to imagine something intermediate between monkey and man. This slot was waiting for apes.

It is customary to say that the propounding of the theory of evolution, that is of the genetic continuity between man and beast, was another great affront to human dignity. If it was so, it did not go deep. For one thing, evolution was already a familiar concept, already widely discussed by philosophers, writing men, and intellectuals. After all, it was only the great ladder of being that laid down along the time coordinate which is traditionally horizontal.

In any case the affront was not so much to people as to Holy Writ and this was soon got over: it came to be seen that the evolutionary concept was as splendid and dignified and as worthy of being a disposition of God's as the doctrine of special creation, and in any case Man was top animal on the evolutionary scale of variation.

I turn now to considering the prolongation of life by attempting to lengthen the lifespan or by the use of medical engineering. Because he was a deeply religious man, I think it specially significant that Francis Bacon in his *Valerius Terminus* describes the true purpose of science as the discovery of all operations and possibilities of operations from *immortality* (if it were possible) to the humblest mechanical practice.

Today research on the lengthening of the lifespan is looked down upon as irreverent and antisocial—antisocial because such a procedure would compound the population problem and to some extent the problem of unemployment. The prolongation of life by something like ten to twenty-five per cent is beginning to be a serious possibility. The life expectancy of mice can be increased by about a quarter by systemic administration of relatively high doses of industrial antioxidants such as 2-mercaptoethylamine. The interpretation is not yet quite certain for these antioxidants are somewhat toxic and they may prolong life by reducing food intake (a procedure already well known to lengthen life). But supposing they do not act this way; suppose we take the results at their face value. If it is true that systemic doses of reducing agents are beneficial this provides a theoretical foundation for the use, energetically advocated by Linus Pauling and Albert Szent-Gyorgy, of high doses of oral ascorbic acid (vitamin C).

The fear that it is undignified to attempt to prolong life in this way will not bear examination. *All* medical treatment, even treatments as humble and simple as indigestion pills and plasters such as we apply to cuts—all such treatments if they are effective at all are effective in a way which on an epidemiological scale can be measured by an increase of life expectancy, however minute it may be.

A more lively topic of current discussion is perhaps the use of medical engineering to try to save victims of physical or medical accidents from what would otherwise be certain death. By 'medical engineering' I mean all the apparatus of intensive care, intravenous feeding, perhaps dialysis as a substitute for kidney function, blood transfusion if necessary, and the mechanical ventilation of the lungs. All this apparatus has been alleged to deprive death of its dignity: it is a 'prolongation of death', the critics say rather than a prolongation of life.

I disagree strongly with this view which is deeply unbiological: there is no more deep-seated biological instinct than that which expresss itself as a firm grasp upon life, there is more dignity, as there is more humanity, in fighting for life than in a passive abdication from our most hardly won and most deeply prized possession.*

* See Chapter 23.

4

Some Reflections on Science
and Civilization
(1972)

At present we are going through a bad period in the relations between science and civilized society. People generally have become increasingly aware and resentful of the havoc that may be wrought by a technology that develops without social censorship. Worse still, a new mood of despondency and helplessness has descended upon us, for the equal of which we should have to go back to the early days of the seventeenth century.

It has come to be felt that there is some *essential* malefaction about the progress of science and technology: that they lurch forward like some great Behemoth trampling down in its pathway almost everything that makes for civilized life. I shall discuss some of the psychological elements that enter into the present tendency to repudiate science and all its works, and then discuss the shape of possible solutions to these embarrassments.

Relations between Science and Society in the Seventeenth Century

If we say that the relations between science and society are now going through a bad period, we are under an obligation to say when we think the relationship was a cordial one. The first period in which it was so was in the 1620s when Francis Bacon most winningly and indeed irresistibly talked people into believing that science was something that would work for 'the merit and emolument of man'. The second, surely, was during the nineteenth century—the epoch of the great civil engineers when it was taken altogether for granted that science was the principal agency of progress and that science and civilization stood shoulder to shoulder. But even then there were rancorous voices; the proposal in 1848 to lay London's first main sewer was vehemently opposed by much established opinion, notably by *The Times* which appeared to feel that to die in one's own way without civil interference was one of the elementary liberties of mankind: 'We prefer to take our chance with cholera and the rest than to be bullied into health. England wants to be clean but not to be cleaned by Chadwick.' (We can take it that the Prince Consort, who died of typhoid in 1861, did not share this bizarre preference.)

Let us now consider some of the psychological elements that enter into the present anti-scientific mood—we say 'psychological' to distinguish them from the sense of resentment people justifiably feel about the ecological depredations of modern technology. The first we shall call 'Mumfordry', by which we mean nothing more serious than a certain propensity to think the worst about science and technology.

In an article published in 1966 Lewis Mumford* quoted a lengthy passage from Edward McCurdy's translation of Leonardo's note-books, a passage which describes the fearful and irresistible rampaging of a black-faced monster with swollen and bloodshot eyes and ghastly features—a monster from which the populace defended themselves in vain:

Oh, wretched folk for you there avail not the impregnable fortresses nor the lofty walls of your cities nor the being together in great numbers nor your houses or palaces! There remained not any place unless it were the tiny holes and subterranean caverns where after the manner of crabs and crickets and creatures like these you might find safety and a means of escape. Oh, how many wretched mothers and fathers were deprived of their children, how many unhappy women were deprived of their companions! In truth . . . I do not believe that ever since the world was created has there been witnessed such lamentation and wailing of people accompanied by so great terror. In truth the human species is in such a plight that has need to envy every other race of creatures.

I know not what to do or say for everywhere I seem to find myself

* *New York Review of Books*, 29 Dec. 1966.

swimming with bent head within the mighty throat and remaining indistin-
guishable in death buried within the huge belly.

Lewis Mumford surmises that this nightmare represents the 'reverse
side of Leonardo's hopeful anticipations of the future'. It is Leonardo's
premonition of the despoliation that the advances of modern science
and technology may wreak upon the earth and its inhabitants. But
'there is no way of proving this', Mumford concedes, and we
therefore diffidently submit an alternative interpretation: that in this
passage Leonardo is describing the devastation consequent upon the
spread of a bubonic plague of Levantine or, perhaps, Libyan origin.
For justice's sake it must be added that the advances of science
and technology have virtually eliminated bubonic plague from the
western world.

The kind of attitude which Mumford's interpretation of this passage
from Leonardo illustrates can only be sustained if we systematically
neglect the benefactions of science—for example, being alive and
well instead of ill or dead—and pay attention only to the miscarriages
of technology. Advances in medicine and the possibilities of human
happiness created by the relief of suffering are a great embarrassment to those determined to think nothing but evil of science and
technology. Their only recourse is to point to the population problem
as the direct consequence of medicine and medical technology and to
say or imply that modern drugs cause as many ailments as they cure.
In spite of these dissonant voices, most people believe as we do that
medical science has a moral credit balance.

Science and the Mastery of Nature

A second element in the revulsion of many thoughtful people from
science—and a factor which has had a thoroughly mischievous effect
for reasons we shall try to explain—is the ideology of the mastery
or domination of Nature. 'Bacon's fundamental maxim' said an
eighteenth-century philosopher 'is that knowledge is power: every
accession man gains to his knowledge is an accession to his power;
and extends the limits of his empire over the world which he
inhabits.' There are passages in Bacon which fully bear out this

judgement. However, in the context of Bacon's own thought the ideology of mastery and domination is not particularly offensive because it is bound up with his own special concept of experimentation as an alternative to the passive contemplation of the information that nature spontaneously proffers us. The idea of mastery makes a brief appearance in the writing of Marx and of Freud. The 'conquest of infectious disease' is fair enough, because bacteria are indeed inimical to us. On the other hand the 'conquest of space' rings quite false. The main objection to the ideology of mastery and warfare is that it dulls the sensibilities and seems to condone or in some perverse way even to justify the worst excesses of environmental despoliation. One hopes that these unpleasant figures of speech have not taken so deep a hold that they cannot be uprooted from popular writing and popular thought. It is *understanding*, not mastery, that should be the declared ambition of scientific research.

Does Science Pay?

Another factor which makes for uneasy relations between technology and civilized life is the tendency, especially prevalent in England, to make economic return or cost-effectiveness the ultimate measure of the worthwhileness of any enterprise. A purely economic system of mensuration is not conducive to the welfare of the environment. The purification and the safe disposal of toxic effluents are costly obligations and the cost-effectiveness of many manufacturing procedures is greatly improved when these obligations are disregarded or circumvented.

Is Society Sick?

Some thinkers of the twentieth century have professed to discern a serious social ailment in modern civilized society. Both Keynes and Freud have had civilization on the couch. J. M. Keynes* diagnosed 'a nervous breakdown' and Freud 'a neurosis'. Keynes believed of mankind that we had been 'expressly evolved by Nature with all

* *The Nation and Athenaeum*, 18 Oct. 1930.

our impulses and deepest instincts' for the purpose of solving the economic problem, the problem of keeping our heads above water in a hostile and competitive world. The economic problem was approaching solution, so Keynes goes on to say, 'I think with dread of the readjustment of the habits and instincts of the ordinary man bred into him for countless generations that he may be asked to discard within a few decades. . . . Must we not expect a general nervous breakdown?' Keynes likens the plight of civilized society to that of a suburban housewife who has been relieved of economic and of some domestic cares by a prosperous husband and for whom the enjoyment of a nervous ailment is now the only remaining *métier*.

Keynes's argument is a relic—the last, one hopes—of social Darwinism and Freud's is based on a misunderstanding, forgivable because at that time widely prevalent, of the true scope and significance of Haeckel's recapitulation theory. Both arguments are unconvincing. The idea that civilized society can suffer from an ailment is a figure of speech which can easily lead us astray. It is fair enough to speak of society suffering from a 'disorder' because orderliness is one of the defining characteristics of a civilized society but we shall depress ourselves unduly if we see ourselves as sitting anxiously by civilization's bedside. It really makes no more sense to speak of civilization suffering a nervous breakdown than to speak of it having a headache or a bad back. The disorders of society are peculiar to and distinctive of society, e.g. an unfavourable trade balance, unemployment, or over-population. They are particular disorders, moreover, for which we should seek particular remedies. The misuse of the organismic conception of society must bear some part of the blame for the current feeling of helplessness and despondency about how matters can be remedied.

Utopia and Arcadia

Let us now consider the shape of possible solutions of our present dilemma, which is that we have become increasingly dependent on science and increasingly resentful of the fact that we are so. In the early seventeenth century, when people in England got fed up with the state of the nation and the quality of life they could and did emigrate to America, but no one today would have recourse to such a solution because America is in a worse plight than we are in most matters to do with the bad relations between science and civilized life.

From the sixteenth century onwards, however, imaginative writers and philosophers have entertained purely notional solutions of the dilemma. These notional solutions or daydreams of a better world are roughly speaking of two kinds: Utopian and Arcadian.

The old Utopias—Utopia itself, the new Atlantis, Christianopolis, and the City of the Sun—were ectopic civilizations discovered by chance in far-off seas. In Utopia man has become the landlord of the domain in which he was formerly a mere tenant. He arranges matters in such a way that people live in amity and under the protection of justice and enjoy all the benefactions that an enlightened technology can bestow upon them—though a technology still so rudimentary that it could not have the grievous side-effects we now deplore. It is worth remembering that the old Utopias did not repudiate science and technology but on the contrary put them to work for the common good. In Utopia it is assumed that men improve themselves by their own exertions and that the qualities of character and mind that make for civilized living will eventually become second nature.

Scientists are Utopian by temperament. If asked why they do what they are doing one suspects most of them would answer that they hoped their work would one day make the world a better place to live in. And that is the essence of Utopianism. A civilized world is not 'given', but something men can make for themselves—an audacious and irreverent idea in its time.

Arcadian thought is closely bound up with the legend of a Golden Age in which human beings lived in a state of natural happiness and innocence, tranquillity and peace—a state from which they have since undergone a grievous decline. In Arcadia human beings retreat into a tranquil pastoral world where peace of mind is not threatened, intellectual aspiration is not called for, and virtue is not at risk. But an Arcadian society is anarchic (a fatal objection). Everything that is implied by authority is replaced by everything that is implied by fraternity. Arcadia is a world without ambition and without accomplishment. Anarchy is perhaps the sociological equivalent of solipsism. Just as it is normally regarded as fatal to the pretensions of any theory of knowledge to show that it leads to solipsism, so, it seems, is it fatal to the pretensions of any social system to show that it leads to or implies a state of anarchy.

Practical Remedies

Let us therefore turn away from notional solutions to practicalities. If we can agree that civilized society is not the victim of any one ailment, we shall not look for any one remedy. It is not one thing that is wrong with modern society but a multitude of particular things and for these we must find a multitude of particular remedies. Pollution, for example, is always pollution of something by something at some particular time and place. Each individual contribution to pollution must be sought out and remedied. If a town is intoxicating itself with smoke then smokeless fuel must be introduced borough by borough until the nuisance has abated.

Sweden has set a notable example of 'piecemeal social engineering' in its conservation programme for Lake Vättern, which was being slowly polluted out of existence and rendered progressively more unfit to supply water to the municipalities round its shore or to sustain a char fishing industry. The major sources of pollution were identified as sewage from holiday camps and lakeshore holiday homes, fermentation liquids from silos and effluents from the regional paper and pulp industry. At considerable expense each source of contamination was studied and as far as possible remedied.

It is a heartening story because it shows how evils of technological origin can be mitigated by technological means. This applies even to the greatest evil of all, the overpopulation problem. We believe that one day a medically inoffensive means will be found of preventing conception and that in spite of administrative difficulties and doctrinal barriers it will come into general use. Nevertheless, our entire economic system, based as it is upon over-production, over-consumption, and waste is unsound and must eventually be supplanted by something more like a spaceship economy distinguished by frugality, recycling, and above all forethought.

The Future

We conclude with a statement of belief which, like all such statements, is essentially a fragment of autobiography.

We believe that technological remedies can be found for evils of technological origin and are prepared to marvel at people who think otherwise. One had hoped that a journey to the far side of the moon had convinced everyone that any accomplishment which is not at odds with the laws of physics is within human capability. (It was some such thought I had in mind when I chose a Baconian motto for my Presidential Address to the British Association in 1969: 'On "The Effecting of All Things Possible" '.*) As to the anti-scientific mood, we can only hope that it will disappear along with the grounds that justify it, unless indeed its causes lie deeper and are less easily eradicable than we have supposed them to be. But of course nothing will happen unless there is widespread public determination that technology must remain subject to rules of social morality and be kept under constant surveillance so that its miscarriages can be prevented or cured. There is already evidence that this new mood of determination is growing. Last year was European Conservation Year. In England a Ministry for the Environment is a promising beginning; the United States has recently held a nationwide Earth Day and the World Health Organization is convening a conference in Stockholm in 1972 on the general theme of man and his environment.

I have said nothing about incentives. There is a widespread feeling that just as politicians and schoolteachers of the wrong kind have found it fairly easy to inculcate ideas of nationalism and patriotism in the narrow sense into schoolchildren, so they might now make amends by inculcating the same sentiments about the earth as a whole. What should the new *ism* be called—terrestrialism? (but that is too much of a mouthful)—and in what terms should children henceforward be encouraged to declare their allegiance to the world they are to live in?

* Published in *Pluto's Republic* (Oxford University Press, 1982).

5

Florey Story
(1979)

Howard Walter Florey was a great man and no mistake. He devoted
the more important part of his professional life to a single wholly
admirable purpose which he pursued until he achieved it, showing,
in spite of many setbacks and rebuffs, the magnanimity that is
the minimal entry qualification for being considered 'great'. In a
memorial address, Patrick Blackett likened Florey's achievement to
that of Jenner, Pasteur, and Lister: but the public were so little aware
of him that when Macfarlane first approached publishers with the
notion of a biography,* they wondered if he would not do better to
write on Alexander Fleming instead. This, Macfarlane surmises,
was because the public had already cast Fleming as the hero of the
great penicillin story: he was a closer approximation than Howard
Florey to the public's stereotype of a great scientist, for, although a
great scientist, Florey was the kind of man who would have been a
success at anything he had chosen to turn his hand to. Macfarlane
thinks the comparison of Florey with Jenner, Pasteur, and Lister is
specially apt because 'the work of these three men forms a logical
sequence with his own that spans, in the course of about 150 years,
the gulf between almost total therapeutic helplessness and the virtual
defeat of most of the important bacterial diseases.' Whatever the
general public may have thought about him, Florey stood unsur-
passably high in the estimation of his colleagues—that which means
most to a scientist—and in due course they elected him head of their
profession in England as President of the Royal Society.

* Gwyn Macfarlane, *Howard Florey: The Making of a Scientist* (Oxford University
Press, 1979).

It is fortunate for mankind that no Geneva convention prohibits the prosecution of germ warfare by germs themselves, among whom the struggle for existence is murderous and unremitting. Penicillin is one of a class of substances manufactured by moulds and bacteria—particularly soil bacteria, which live in deplorable conditions of squalor and overcrowding: these are substances which suppress the growth or multiplication of other microorganisms. Penicillin was discovered by Alexander Fleming: by luck, so it is believed, though in reality Fleming had been looking for something very like penicillin all his life. The only element of blind luck about the discovery of penicillin was that, unlike most antibiotics, penicillin is not poisonous to human beings and other higher animals: the reason is that penicillin interferes with metabolic processes peculiar to bacteria, whereas some other antibiotics like actinomycin and mitomycin are toxic because they obstruct a cellular activity common to bacteria and ourselves.

Before a good scientist tries to persuade others that he is on to something good, he must first convince himself. The first experiment that convinced Florey and his two colleagues, Norman Heatley and Ernst Chain, that they might be on to something occurred very shortly after the German Army—'to the inexplicable surprise of the Allied Command—instead of dashing themselves to pieces on the Maginot line, drove their Panzer columns round the northern end of it, and swept between the British and French armies against almost no resistance until they reached the coast near Abbeville'. At 11 a.m. on Saturday, 25 May 1940, eight white mice received approximately eight times the minimal lethal number of streptococci. Four of these were set aside as controls, but four others received injections of penicillin—either a single injection of 10 milligrams or repeated injections of 5 milligrams.

The mice were watched all night (but of course). All four mice unprotected by penicillin had died by 3.30 a.m. Heatley recorded the details and cycled home in the black-out. Poor mice? Yes of *course* poor mice, but poor human beings too, don't forget:

Next morning, Sunday 26th May, Florey came into the department to discover that the results of his experiment were clear-cut indeed. All four control mice were dead. Three of the treated mice were perfectly well; the fourth was not so well—though it survived for another two days. Chain arrived, and then Heatley, who had had very little sleep. They all recognised that this was a momentous occasion. What they said is not recorded, but memory has supplied subsequent writers with various versions. One might suppose that Heatley said very little, that Chain was excited, and that Florey's reported comment 'It looks quite promising' would be entirely in character.

Animal experiments on a much larger scale soon made it clear that penicillin was indeed of great potential importance. The first published paper on the subject in the *Lancet*, by Chain, Florey, Gardner, Heatley, Jennings, Orr-Ewing, and Sanders—the names are in alphabetical order—stated: 'The results are clear-cut and show that penicillin is active *in vivo* against at least three of the organisms inhibited *in vitro*.'

Macfarlane's account of the animal experiments and the first clinical trial is simple and straightforward, and all the more exciting for being so. It makes my heart pound still although I know the outcome, for the thrill of reading about these great occasions does not diminish: scientists are like cricket-lovers who never tire of reading or recounting what Colin Milburn used to do to short balls on the leg side from Wesley Hall.

Florey's and his colleagues' clear awareness of its importance raised the problem of what would become of their strain of *Penicillium* if—as seemed entirely on the cards—the German Army were to reach Oxford and sack it: 'The mould itself must be preserved, undetected. Florey, Heatley, and one or two others smeared the spores of their strain of *Penicillium notatum* into the linings of their ordinary clothes where it would remain dormant but alive for years.'

Because of its potency and non-toxicity penicillin is the paradigm of antibacterial substances, but it is not without snags: the warfare between germs which, as I suggested above, leads to the formation of substances like penicillin leads also to the evolution of remarkably effective mechanisms of defence. One is the manufacture of the ferment *penicillinase* which destroys penicillin and thus protects bacteria from it. The widespread use of penicillin—sometimes injudiciously often—has led to the evolution in many hospitals of strains of bacteria resistant to its action: once a mutant impervious to penicillin has arisen, natural selection soon brings it about that the mutant becomes the prevailing type in the population. It is not that penicillin has lost any virtue, but rather that bacteria have acquired a vice.

Another snag, exacerbated by the tendency of clinicians in the early days of penicillin to administer colossal intramuscular doses in the presence of substances known to immunologists as 'adjuvants', is that penicillin can give rise to severe allergic reactions in a specially susceptible minority of those who receive it. The development of new antibiotics or new variants of penicillin has gone a long way to annul this disadvantage.

Among the difficulties associated with the production and use of penicillin, I have mentioned only the evolution of penicillin-resistant strains of bacteria, and the ability of penicillin to sensitize susceptible patients. The greatest snag of all was the sheer difficulty of producing quantities sufficient for clinical trials. But Florey's School of Pathology became a pilot-scale production plant for the purpose, and it was natural that he should turn to the great pharmaceutical companies to make use of their great practical experience and know-how. One of the American companies, Merck, knew how only too well. According to Macfarlane's account, Norman Heatley's visit to share with them what he knew about the production of penicillin was marked by something much less than candour on the part of Merck, who had prepared applications for British and American patents covering the essential stages of production processes devised by the Oxford scientists and one of their own scientific officers: 'This fact was not generally appreciated until 1945, when British firms discovered that they had to pay royalties on their penicillin production.' Whatever their moral shortcomings, the great pharmaceutical companies did in due course produce penicillin in adequate amounts— the consideration that mattered most. The complexity of the production of penicillin and the murkiness of its origin ('see *Macbeth*, Act Four, Scene One,' said Oxford wags not unwilling to poke fun at a discovery so obviously important) impeded the funding of Florey's research, for in the 1930s Gerhard Domagk's discoveries had ushered in the era of the sulphonamides, also powerful antibacterial agents. They, being synthetic organic chemicals, could be produced without cauldrons and the toil and trouble that go with them. It is clear that some know-alls serving on the Medical Research Council at the time must have resolved that the future of antibacterial therapy lay with these synthetic organic chemicals and not with 'biologicals' such as penicillin, for, much to the annoyance of Florey and Chain, the MRC did not fund penicillin as handsomely as the occasion called for. Luckily, however, the Rockefeller Foundation helped out. The sums involved—of the order of hundreds of pounds—seem comically small by modern standards, but money went much further then.

A leading article in *The Times* on 'Penicillium' referred, without mentioning any names, to the research in progress in Oxford. Sir Almroth ('stimulate the phagocytes!') Wright addressed the Editor thus about his former pupil:

Sir,
In the leading article on penicillin in your issue yesterday you refrained from putting the laurel wreath for this discovery round anyone's brow. I would, with your permission, supplement your article by pointing out that, on the principle *palmam qui meruit ferat,* it should be decreed to Professor Alexander Fleming of this laboratory [St Mary's Hospital]. For he is the discoverer of penicillin and was the author also of the original suggestion that this substance might prove to have important applications in medicine.

Ever since Wright's letter, there have been attempts to make a *cause* out of the allocation of credit for the great discovery: first by comparing the contributions of Fleming and Florey, and more recently by comparing the contributions of Florey and Sir Ernst Chain. But no journalist will get any copy out of Macfarlane: he treats the whole subject wisely and temperately, as might be expected of a historian who is a distinguished scientist.

Human nature, unfortunately, is such that so great a discovery as that in which Fleming and Florey played crucially important parts is certain to be followed by jealous attempts to diminish it by finding evidence that it had all been thought of or done before. Certainly Pasteur recognized that germs engaged in germ warfare, and may be the Chinese *did* put mouldy soya bean curds to therapeutic uses: but Alexander Fleming discovered penicillin, and Howard Florey was the prime mover in turning it into the most important therapeutic innovation of the twentieth century. Both were necessary, but neither can be judged singly sufficient.

Relations between Florey's team and Fleming were inevitably difficult. Florey did what was proper: that is to say, he acknowledged Fleming as *the* discoverer of penicillin in his first paper on the subject, but Fleming always felt he deserved more credit than he got, and referred often to 'my brainchild'. 'What have you been doing with my old penicillin?' Fleming asked the Oxford team when he came down to visit Florey's laboratory. Florey and Chain told him what they were doing and took him on a tour of the laboratory: 'Fleming said almost nothing during this inspection and returned to London without comment or congratulation on what had been achieved.' It is clear from Ronald Hare's *The Discovery of Penicillin* (1970), upon which Macfarlane draws gratefully, that Fleming was an amateur in the big business of practical therapeutics, where Florey was every inch a pro.

Florey was the greatest experimental pathologist of his day. Penicillin was not his only—nor even his principal—interest: to judge by the quietly passionate persistence with which he studied the problem and persuaded all his young colleagues to do so too, the central interest of his scientific maturity was to elucidate what came to be called 'the great lymphocyte mystery'. Lymphocytes are white blood corpuscles—those, as we now know, that transact immuno-logical reactions. The lymphatic vessels of the body, which drain fluid from the tissues, unite into one major vessel, the thoracic duct, that empties its contents directly into the bloodstream. By this route thousands of millions of lymphocytes enter the bloodstream daily, but what becomes of them, and what is their function anyway? Whatever their intentions may have been, most of Florey's young students and co-workers—among them a future Dean of the Harvard Medical School, a future head of the Medical Research Council, and the future wife of the reviewer—found themselves trying to answer these questions. The problem was solved by Dr J. L. Gowans, the one who became head of the MRC: he found, contrary to orthodox opinion, that lymphocytes are relatively long-lived cells which circulate and re-circulate through the blood and lymph vessels. Like the chorus in a provincial production of *Faust*, lymphocytes in the bloodstream at any one moment disappear behind the scenes and re-enter by another route. Lymphocytes, moreover, are cells that manufacture antibodies and are responsible for the recognition, and ultimately the elimination, of foreign-tissue transplants and cells infected by viruses or transformed by the action of cancer-producing agents.

Another of Florey's great interests was the nature and cause of atheromatosis, the formation in blood vessels of atheromata, the waxy plaques that sometimes threaten the free passage of blood in such important vessels as the coronary arteries, which supply blood to the great muscles of the heart..

As a man of action, determined to get results, Florey might easily have made some bitter enemies, but in reality he inspired a good deal of affection and admiration among colleagues, who to this day like exchanging Florey stories and laughing at their discomfiture when some characteristically sardonic or scathing remark of his cut them down to size. I wrote a long-winded paper on my work as a graduate student in his lab and showed it to Florey. When he handed it back to me, he said: 'I don't see what you're getting at, Medawar. The paper doesn't make sense to me.' Later on, having learned better, I wrote a clear and simple paper for a journal Florey was particularly fond of and regularly read, and I was overjoyed when Florey passed me in one of the narrow lanes that wind through the science area in Oxford, twitted me in his usual style on having rushed precipitately into print, but added: 'Your paper's not at all bad, Medawar.'

Florey was extremely intelligent, clear-sighted and shrewd, but he was not an intellectual, and even at the height of his success he was a tiny bit scared of such people, for he was not always as sure of him-self as might have been expected of a man so enormously successful. 'Was that all right?' Florey once said to me after giving an important lecture at the Royal Society, of which he was President. It was, but I thought it touching and endearing that he should still want to be assured of the good opinion of his juniors.

So deep was the impression made by Florey on his juniors that I do not believe any one of them could have written a life of Florey of which Florey would himself have disapproved. He would have liked Macfarlane's life because it is simple and straightforward and sticks to the point without clever philosophical or psychological digressions. Florey might have made some amusedly self-depreciatory remark about the use of the word 'great', but if he had done so he would have been—as he seldom was—wrong.

6

The 'Ultra-Élite' of Science
(1977)

The predominant mood of the world today is such as to make 'élite' a dirty word, but the only concession Professor Harriet Zuckerman* makes to the politics of envy is to quote *in extenso* a laboriously argued passage from Pareto which seems to authenticate the use of the notion in sociology.

Four hundred and ninety-three thousand citizens of the United States described themselves as 'scientists' when asked their occupation in the national census. Of these, 313,000 were recognized as such by the taxonomic criteria adopted by the National Science Foundation. In 1976, 77 Nobel laureates were living in the United States. This represents 1/7000 of self-defined scientists and 1/4000 of scientists defined by the criteria of the National Science Foundation.

Professor Zuckerman regards the Nobel laureates as an 'ultra-élite' of science, but of course numerical rarity is not enough, and one is tempted to ask if an élite could be so described if it were not popularly regarded as such. The Copley medal, the highest award of the Royal Society of London—the most famous and for all practical purposes the oldest scientific society in the world—is a rarer and, it can be argued, an even higher mark of distinction than the Nobel prize, but as no one outside the Royal Society, and not everyone in it, thinks so, the Copley award fails to define an élite. Yet no one other than a superlatively good scientist has ever won the Copley medal and some have it who did not, but should have received a Nobel award, notably O. T. Avery and N. W. Pirie. Moreover, the character of the Copley medal is such that it can be awarded to

* Harriet Zuckerman, *Scientific Elite: Nobel Laureates in the United States* (Collier, 1977).

scientists who are not eligible for Nobel awards but are of the stature of Nobel laureates. I think particularly of R. A. Fisher, the geneticist, and G. H. Hardy, the mathematician.

Rather similar considerations apply, in the United States, to membership of the National Academy of Sciences—the American equivalent of the Royal Society—a high distinction certainly, but not one which is held in awe by the general public nor even one that is kept consciously in mind as a goal of endeavour by the scientific profession—as a fellowship of the Royal Society of London certainly is.

I myself am envious that this polished piece of research was not carried out on the fellowship of the Royal Society in the United Kingdom. If the author had had nothing but the interests of scientific research in mind, it would have been a better choice than that of Nobel laureates in the United States. I earnestly wish she would one day come over here and study us.

The Royal Society is an institution which every scientist wants to belong to—so badly, sometimes, that failure to be elected may cause deep distress. Julian Huxley once told me of a colleague who, when the news had to be broken to him that his name was not among those nominated for election in one year, burst into tears in Paddington station. I am happy to say that he did eventually get in; happy because one cannot but deplore the distress which is caused by the sometimes faulty working of a fallible electoral system. The Royal Society owes its success to being just the right size: small enough for election to be a distinction, but large enough for every first-rate scientist in the Commonwealth to hope for admission. The very greatest scientists lived before Alfred Nobel got the knack of stabilizing the nitric acid esters of polyhydric alcohols— notably trinitroglycerol—but a fellow of the Royal Society still exults in being one of the company to which Robert Hooke, Robert Boyle, Christopher Wren, Isaac Newton, Michael Faraday, Benjamin Franklin, Charles Darwin, Josiah Willard Gibbs, and James Clerk Maxwell also belonged. However that may be, this book is not about the Royal Society but about American Nobel prize-winners.

For her field work Professor Zuckerman interviewed 41 of the 56 laureates at that time in the United States and tape-recorded all but one of them. The choice of prize-winners turned out to be wise, for 'the laureates were accustomed to talking about themselves and their work to visiting outsiders'. I hope Professor Zuckerman did not suffer too grievously for science: I know one laureate who has no topic of conversation other than the marvel of his laurel headgear— a topic to which his conversation invariably returns. The 41 laureates interviewed were not the whole population from which she drew information; her study is based on 92 laureates selected between 1901 and 1972 who worked in the United States on the research which was to bring them their prizes.

Professor Zuckerman is a sociologist and although this book is avowedly a piece of sociological research it has none of the characteristics which have rightly brought so much sociological research into public ridicule; there are no polysyllabic words of the kind used by Victorian biologists to try to impress the lay public with their profundity, and the writing throughout is clear and straightforward—unmistakably the work of someone who has something to say and knows how to say it. It is indeed a well-executed piece of scientific research of which the subject happens to be an aspect of human behaviour—an ethology practised upon society rather than an individual organism.

Readers used to the tempo and idioms of feature journalism, particularly in magazines with many coloured illustrations, will probably find Professor Zuckerman's book too slow-moving for their debauched tastes, but that is only because they do not understand that an important part of the function of social science in such a context as this is either to authenticate or finally to discredit the kind of 'pop sociology' which is heard in everyday speech. My favourite is the kind of brash declaration which can be heard in the bar of almost any suburban tennis club populated by young executives: 'Of course, the trouble with the Japanese is that they have no real inventive powers, they can only imitate others.' I have heard this with my own ears at a time when it should already have been clear even to the meanest intelligence that Japanese scientists and technologists are enormously inventive and imaginative as well as inexhaustibly ingenious; and that their very common tendency to allow other nations to bear the costs of research and development, while they apply their inventive skill to improvement and further development, is a trait which does great credit to their common sense and business acumen.

I am afraid, though, that serious social scientists such as Professor Zuckerman will long have to put up with the kind of uncomprehending ridicule which pioneer scientists of the Royal Society had to put up with from Charles II and his courtiers, who thought it irresistibly comic that scientists should try to weigh air. Sociologists will survive, though, just as scientists did, if they cleave to the standards of scholarship which are apparent throughout Professor Zuckerman's book.

Professor Zuckerman deals in some detail with the character of the research which was eventually thought worthy of receiving the Nobel prize, and in passing she observes that 'seniority without prime research achievement is not enough'. The same is true of election into the Royal Society, many of the candidates for which feel that many patient years of hard worthy labour should he rewarded by election. But as a rule it is not; there must be some tinkling of bells before a candidate will be considered seriously.

Some of Professor Zuckerman's best writing is on the question of whether or not important scientific innovations are characteristically the work of the young. Anecdotal evidence is rightly regarded as insufficient, and she points out—what is obvious only to those with some understanding of demography—that one cannot appraise evidence of age in scientific discovery without taking into account the age distribution of the population 'at risk', as actuaries say—at risk, that is, of making a contribution to science. Scientists have been the subjects of a kind of population explosion of their own. Like the population generally, the age-distribution has a youthful pattern. However that may be, the average ages at which laureates did the work which subsequently won them their Nobel prizes follows the expected pattern: the physicists are youngest, then the chemists, then the prize-winners for physiology or medicine. The average age for all fields is only 39·2. The average age at which they won their prizes was 52.

From the standpoint of scientific administration, no sociological research could be more important than the attempt to ascertain what truth there may be in the commonplace notion that creativity diminishes with advancing years. On such a subject as this it is almost impossible to avoid citing what medical men call 'clinical impressions' and to forbear from calling upon the Verdi of *Falstaff* to take the witness stand. Sometimes, of course, there is a distressingly long delay between the execution of prize-winning research and its reward. Peyton Rous discovered the propagation of chicken tumours by viruses quite early in the century, but was 87 before he received his Nobel prize.

Scientists and mathematicians who do great things when very young do not always fulfil their early promise, but those who die young are always given the benefit of any doubt and live for ever more on the most brightly sunlit slopes of Parnassus. Roger Cotes is one such and the mathematical logician Frank Ramsey another.

When, as in actuarial practice, a proper allowance is made for the age-distribution of the potential candidature for Nobel prizes, the results are rather surprising:

If anything, it is not the young who turn up disproportionately often among those making prize-winning contributions, but the middle-aged: 23 per cent of the laureates were 40 to 44 years old when they did their prize-winning research, but only 14 per cent of the run of scientists fall into this age cohort. (p. 168)

As to where the prize-winning research was done, thirty scholarly institutions account for them all, and six for somewhat more than half of the Nobel prizes awarded to Americans. The Rockefeller Institute and Foundation score eight, and it is a sign of the times that the Bell Laboratories score four laureates in physics. Among American universities the score in terms of prizes won runs in the order Harvard, Columbia, California (Berkeley), Washington, Stanford, Cornell, Illinois. A governmental research institution, the National Institutes of Health, occupies about a half-way position. It is difficult to record without a note of smugness that the National Institutes of Health score (3) falls far short of that which can be credited to the British Medical Research Council (10). I think this may point to a genuine difference in the degree of esteem in which governmental research establishments are held in the United Kingdom and the United States.

Professor Zuckerman writes amusingly about how learned institutions compete for the credit of owning a Nobel prize-winner: '. . . the College of the City of New York has given a prize to Arthur Kornberg its first alumnus to win the Nobel—perhaps the first time someone has won an award for having won an award.' I was a little surprised by this last comment, because I believe it to be more often true that one gets one award for having got another. Certainly whenever I have received an honorary degree a public orator has always given it as a good reason that I have had honorary degrees from so many other universities. Here is an interesting nugget of information:

A few institutions observe self-denying ordinances that limit their claims to laureates. Harvard does not list its laureate alumni, nor does it count those who were on its faculty at some time or another before or after winning the Nobel prize. It claims only those who did their prize-winning research at Harvard and those who were on its faculty when the award was made. (p. 31)

This consciously high-minded repudiation of the 'George Washington slept here' attitude of more newly founded institutions may be interpreted either as magnanimity or as a deft piece of one-upmanship. I incline to the latter interpretation just as I incline to the view that Jean-Paul Sartre's refusal of a Nobel prize in literature was done in the consciousness that it would not sacrifice *réclame*.

That a class distinction is embodied in the Harvard attitude is made clear by Professor Zuckerman:

high-ranking universities generally confine themselves to mentioning winners of major scientific awards; those at the second and third levels of prestige do not neglect what award winners they have but will also indicate, for example, what faculty members hold office in scientific societies. Those at the lowest level are largely confined to using such indicators as the number of papers published by their faculty members, as the best measure they can muster.

I have a feeling that in relatively small scholarly institutions the Nobel award can have a disruptive effect. When I took my first chair at the University of Birmingham, one member of the staff had won the Nobel prize and his opinions were thereafter held in such respectful awe that his word could outweigh that of the entire faculty; but in spite of the dangers which go with such a position of privilege, I am happy to say that Norman Howarth's authority was, in the main, beneficent.

It is inevitable that there should have been mistakes in the award of Nobel prizes. Every professional scientist can think of errors of omission. I, for example, think it quite amazing that the elucidation of the functions of the thymus gland should not have been thought worthy of an award. One laureate, Fibiger, received an award for what was later officially admitted to be a non-discovery. The ill-judged award of the Nobel prize to the head of the laboratory in which Banting and Best did their famous work on insulin reminded me of J. B. S. Haldane's bitter remark that a high military decoration for research on chemical warfare was awarded to the man who opened the door of the inspecting general's car.

Describing the nomination and election procedure, Professor Zuckerman calls attention to the importance of multiple nominations and mentions also the dismayingly long delay some future laureates have to put up with before their work is recognized. C. S. Sherrington's case is notorious, for although he had been the world's leading neurophysiologist for twenty or thirty years, he did not receive the Nobel prize until 1932. The difficulty, people explain to each other, was that Sherrington did not make one major discovery of the kind which attracts the Nobel award. All he did was to lay the foundations of the reflex physiology of the nervous system. To this very day it can be said that our knowledge of how the nervous system works owes more to Charles Sherrington than to any other single man. For the painfully long delay in the award to Peyton Rous, there is no such excuse. In addition to being the greatest experimental pathologist of his day, he had also made one discovery of quite spectacular importance: the discovery that some tumours are infectious in origin and may be propagated from one animal to another by a virus. A literature citation index—a record of how often a scientist's work is referred to by others—may be used to quantify the scale of influence of a Nobel laureate upon the science of his day. I believe no such record was in existence in Sherrington's day. Had it been so his influence would have been seen to be all-pervasive. Sherrington's work was indeed the foundation of all the great advances in neurophysiology which took place in his day and since then.

'The 41st chair in science'—41 being one more than the number of 'immortals' in the French Academy—is Robert Merton's figure of speech for scientists who ought to have won the Nobel prize but were unluckily crowded out by others. Occupants of this 41st chair acquired an almost official status by their citation in the official history of the Nobel awards. Some such omissions—notably that of O. T. Avery—are so strange as actually to diminish the stature of the Nobel prize. Many others are certainly regrettable: C. R. Harington, E. H. Starling, and Jonas Salk. Sometimes the world of science is itself to blame: Josiah Willard Gibbs, a giant figure in American science along any scale of measurement, was never nominated. Men such as Herbert M. Evans and O. T. Avery were so enormously influential and so highly esteemed that it is hard to see how the Nobel award could have made them more so. Professor Zuckerman's generously lengthy Appendix D is a semi-official list of the occupants of the '41st chair' and there is an explanatory comment about each entry.

Mark 4:25 is pretty explicit on a characteristic which Professor Zuckerman shows to be of special importance in the upbringing of Nobel laureates: 'For he that hath, to him shall be given.' The cumulative character of the advantages they enjoy both before and after their awards is shown not merely by monetary support of research but also by their increased power to attract able colleagues.

There are many political or temperamental reasons why we incline to hold rather definite views on the social provenance of Nobel laureates: for example, that most Nobel laureates are of working-class origin, or are Jews, or come from happy and stable homes. It is in sorting out the rights and wrongs of claims such as these that the professional sociologist shows her strength. Élites in nearly all departments of social life come disproportionately from the middle and upper occupational strata. This is true of American Nobel laureates, too, for 90 per cent of their fathers were professional men or managers or proprietors; the corresponding figures are 74 per cent for high-ranking military or naval officers and 90 per cent for

Supreme Court justices. Even if some laureates are of comparatively humble origin it is pretty clear that being born into an educated and comparatively well-off family helps: 'Even in a system as meritocratic as American science, in which *identified* talent tends to be rewarded on the basis of performance rather than origin the ultra-elite continue to come largely from the middle and upper middle strata.' Whatever the explanation of this may be, whether we look to nature or to nurture, 'the fact itself is clear: the social origins of Nobel laureates remain highly concentrated in families that can provide their offspring with a headstart in access to system-recognized opportunities' (pp. 67–8).

Nobel laureates show the expected partition in respect of religious profession or upbringing—not necessarily religious practice or sincere belief: the Jewish people are relatively abundant while Roman Catholics are disproportionately few. *Émigrés* from Fascist tyranny augmented the numbers if not the proportions of Jewish laureates but the magnitude of Hitler's self-inflicted injury can be gauged by the fact that five of the scientific Jewish emigrants were Nobel laureates already: Einstein, Enrico Fermi, James Franck, Otto Loewi, and Otto Meyerhof. Professor Zuckerman makes special mention of the contribution of the smaller central European states. Forgetting about science for a moment, one cannot but think it noteworthy that England's two greatest and most deeply admired scholars are both by origin Viennese Jews: Ernst H. Gombrich and Karl R. Popper. The Hungarian scholars and artists deserve a sociological treatise for themselves alone. Certainly Leo Szilard, Michael Polanyi, Edward Teller, John von Neumann, Arthur Koestler, Thomas Balogh and Gustav Nossal, and Nicholas Kaldor amount to a rather amazing array of talent. Professor Zuckerman warns us, however, that some caution is necessary in interpreting these numerical data:

In reckoning the extent of the Nazi effect, we cannot indulge in conjectural history and suppose that the young Hitler-emigrés who left Germany and later did prize-winning research would have done work of the same significance had they stayed. Indeed, as more than one said in the course of my interviews with them, having been forced to leave Germany turned out to be the best thing that could have happened to them. The United States provided an attractive and hospitable climate for their work, and for many, ample resources as well. But if emigration was highly beneficial for some of the scientists individually, it was scarcely the best thing that could have happened to German science, and its effect on world science is still not clear. (p. 71).

Other considerations to be borne in mind in evaluating the representation of Jewish scientists among American laureates are the tradition of reverence for learning in Jewish families, the well-known predilection of Jews for professional occupations, and the high proportion of them who enjoy higher education: 'in the early 1970s, for example, 80 per cent of American Jews of college age were in college as compared with 40 per cent for the college-age population as a whole . . .' (p. 71).

Like a true scientist, Professor Zuckermann acknowledges very scrupulously her indebtedness to those upon whose work she has drawn for factual information. In the context of the classification of American university faculty by religious affiliations, this means especially the work of Lipset and Ladd. I was particularly happy to see this evidence of scholarly courtesy because scholars of different disciplines attach very different weights to the propriety of acknowledging the help of their predecessors. Scientists on the whole are good at it, and acknowledgement of indebtedness is very much a part of the scientific ethos: philosophers in my experience are very bad at it—'it is a scandal', a philosopher told me—and very often philosophers write as if in their chosen field of investigation no plough had ever turned the soil before. A great deal depends, I suspect, upon whether or not a 'paper' in a learned journal is the recognized and orthodox means of scholarly communication.

After her satisfyingly thorough investigations into the social and religious provenance of the laureates, Professor Zuckerman turns to consider the relatedness of Nobel laureates. There are some blood relationships—in England, the Braggs *père et fils* won prizes for the same discovery, but the Thompsons for different discoveries.

It is more often a matter of master-apprentice relationship. Sometimes, though, the stable seems to count for more than the pedigree. The old Rockefeller Institute, the great Cavendish Laboratory and the Department of Physiology in Cambridge all have a remarkable record.

As to the relationship of master and pupil, Professor Zuckerman notes that:

it is striking that more than half (forty-eight) of the ninety-two laureates who did their prize-winning research in the United States by 1972 had worked either as students, post-doctorates, or junior collaborators under older Nobel Laureates. (pp. 99–100)

A clearly set out genealogy of master and pupil makes Professor Zuckerman's point perfectly; thus Hans Krebs, one of the world's greatest biochemists, was Otto Warburg's pupil, Emil Fischer's grand-pupil, and Adolph von Baeyer's great-grand-pupil; and this remarkable lineage can be traced still further back, to Kekulé, Liebig, Gay-Lussac, and Berthollet—all household names in science.

Professor Zuckerman does not think that politicking has much to do with the tendency of Nobel prizes to run in such lineages because although all laureates have the right to nominate for awards they are only a small minority of those invited to do so. There need be no one explanation of the phenomenon. The fact that Nobel laureates are highly influential and naturally like to promote the interests of their young must surely be one of the factors at work. Where masters and apprentices have both won the Nobel prize, the modal difference between their ages was 16–20 years. The master is not always the older: Pauli was two years younger than Rabi when Rabi chose to work with him. Detailed and clearly set out tables summarize the relevant information, from which one or two consistent patterns emerge: young laureates-to-be were 26 or 27 when they apprenticed themselves to their respective masters.

Professor Zuckerman's American laureates received their Nobel prizes at the average age of 51; the apprentices, though, received their prizes at the comparatively early mean age of 46, a figure which doubtless contributed to the widespread belief that 'science is a young man's game'.

Among the elements of the heritage which apprentices receive from their masters is the self-confidence which comes with the successful execution of worthwhile work, a sense or 'nose' for a truly worthwhile problem—one really worth what may be the great exertion of trying to solve it. To these we must add the powerful *vis a tergo* imparted by an influential man who would like his pupils to succeed in life; the whole pattern of relationship tends to be self-perpetuating because apprentices when they, in turn, become masters tend to reproduce the style and manners of their own masters.

In describing the history of the Nobel award, Professor Zuckerman makes a point which is often overlooked: it was Nobel's intention that his award should relieve its recipient of want, provide funds for his research, and enable him to devote his whole life to science. These were not unrealistic ambitions when the Nobel prize was started, though today the value of the award is not enormously greater than a year's income of a scientist who is distinguished enough to receive the prize. It certainly would not be enough to keep him and to pay for his research for the rest of his life. In contrast, when I myself began research I had no difficulty in financing my research, living fairly well, and marrying on a salary of £350—which was then less than $2,000; but even though things were very much cheaper in those days, a Nobel prize would have come in very handy. The degree to which the prestige of the Nobel award is used politically can be gauged from the number of manifestos which list Nobel signatories. I do not know whether the Nobel signatories give these manifestos quite the influence which their sponsors imagine, and I have a strong suspicion that a Nobel prize-winner's signature is being progressively devalued by the number of occasions on which it is used. I can remember once being asked to sign a manifesto on something like the following lines: 'The nations of the world must henceforward live together in amity and concord, and abjure the use of war, as a means of settling political differences.' This declaration embodies a splendid ambition, but as one reads it one cannot help wondering if there is any large and influential body of opinion to the contrary. Can we, for example, find a body of men as numerous as the Nobel laureates who believe that the countries of the world should live together in discord and hatred and should have immediate recourse to war when their political interests are in any way in conflict?

One of Professor Zuckerman's most interesting chapters, 'After the Prize', has to do with the impact of award on those who receive it. Alfred Nobel would certainly have been much dismayed if it had occurred to him as a possibility that the benefaction intended to put its recipients beyond the reach of want had, in reality, deprived them of any further incentive to continue with scientific research. The laureates who would have pleased Nobel are those to whom the award gave a terrific filip, making it possible for them to proceed with their research with all the benefits of security and in the warmth of public recognition. Alas, it is not always so, for although I know one laureate who, having completed the work for which he was later awarded the Nobel prize, went on to do research equally worthy of an award, I know one other who has given up research altogether and who travels the world from one conference to another, most of them with titles like 'Man and the Universe' or 'Man and the Future'—conferences which have in common that no one quite knows what they are about—perhaps because they are not about anything. In terms of number of papers published, Professor Zuckerman discerns a certain decline in productivity after the receipt of the Nobel prize, but I wonder if the same would be true if it were possible to give each paper a weight according to its quality and scientific significance. A Nobel laureate is not likely to publish potboilers after his award and, to avoid the suspicion of doing so, may publish rather seldom.

Although laureates often complain about the demands on their time which the award entails, it is not in reality greater than that suffered by any other public figure. Professor Zuckerman quotes *in extenso* Dr Francis Crick's check-list:

> Dr. Crick thanks you for your letter but regrets that he is unable to accept your kind invitation to:

send an autograph	help you in your project
provide a photograph	read your manuscript
cure your disease	deliver a lecture
be interviewed	attend a conference
talk on the radio	act as chairman
appear on TV	become an editor
speak after dinner	write a book
give a testimonial	accept an honorary degree

Dr Crick sounds rather peremptory and ungracious, but this strikes me as a legitimate form of self-defence, the more justifiable in

Crick's case because he was the laureate who may be said to have won the Nobel prize with bar.

Professor Zuckerman writes very understandingly about the problem of allocating priorities in joint work—i.e., of deciding the order in which authors' names shall appear on their joint publications—a matter of some importance in the United States, where the first name is assumed to be that of the major contributor. I myself applaud the Royal Society's alphabetical rule. It may be hard luck on persons with names such as Zuckerman or Zygismondi, and disproportionately felicitous for the Aaronsons of the world, but it is an accepted convention and one which saves much agonizing and resentment. I do not think that there will ever be a general rule for deciding upon which member of a team of two or more was the prime mover or the first to have an idea. My reason for saying so is that scientific colleagues are synergistic—their combined talents and inventiveness are greater than the sum of individual abilities.

As in the gag-writing sessions in which comics try to think up new jokes, each member of the team builds upon the ideas of the other till there is no knowing whose idea the final product was.

Scientific Elite is a standard and definitive work. It need not be done again, but if it is, the new research is almost certain to be avowed either to corroborate or to call attention to the need for qualifying some of Harriet Zuckerman's main conclusions. I guess the former.

7

Scientific Fraud
(1983)

Some policemen are venal; some judges take bribes and deliver verdicts accordingly; there are secret diabolists among men in holy orders and among vice-chancellors are many who believe that most students enjoying higher education would be better-off as gardeners or in the mines; moreover, some scientists fiddle their results or distort the truth for their own benefit.

None of these, though, is representative of his profession—and only people young enough to be cynical believe them to be so. The number of dishonest scientists cannot, of course, be known, but even if they were common enough to justify scary talk of 'tips of icebergs' they have not been so numerous as to prevent science's having become the most successful enterprise (in terms of the fulfilment of declared ambitions) that human beings have ever engaged upon. The profession, sticking together (which is not such a bad thing to do), believes that cheating in science is a curious minor neurosis like cheating at patience—something done to bolster up one's self-esteem. Rather than marvel at, and pull long faces about, the frauds in science that have been uncovered, we should perhaps marvel at the prevalence of, and the importance nowadays attached to, telling the truth—which is something of an innovation in cultural history, if by the truth we mean correspondence with empirical reality. The authors of the more lurid travellers' tales would have been taken aback if someone had described them in modern vernacular as 'bloody liars', but so they were, many of them. They were telling stories, and wanted to tell good stories. Aristotle's conception of poetic truth was one in which correspondence with reality played little part, and his biology gave an account of what he thought *ought* to be true in the light of his deep conception of the true purposes of nature. Thus it ought to be true according to the hebdomadal rule that male semen is infertile between the ages of seven and twenty-one—a pathetic absurdity of which Aristotle would not have been guilty if he had had any real sense of empirical truth. Aristotle was a pioneer, perhaps, in what I believe to be the commonest form of self-deception in science: the kind of attachment to a dearly loved hypothesis that predisposes us (yes, all of us) to attach a special weight to observations that square with and thus uphold our pet hypotheses, while finding reasons for disregarding or attaching little weight to observations and experiments that cast doubt upon them. There is no one who does not roll out the welcome mat with a flourish for those who bring evidence that upholds our favourite preconceptions.

The most puzzling fraud of all—for such in effect it was—was that of the segregation ratios (3:1; 9:3:3:1) as reported by Gregor Mendel in his plant-breeding experiments. As R. A. Fisher was the first to point out, these ratios conformed far too closely to theoretical expectations to be plausible, having regard to the numbers of plants and seeds involved. The explanation could be as simple as that Mendel was a nice chap whom his gardeners and other assistants wanted very much to please, by telling him the answers which they suspected he would dearly like to hear: moreover, as Mendel was an abbé, his assistants may have felt that there was an element of heresy in securing results other than those the Reverend Father was convinced were true. This is a subject on which the authors of the present book* write amusingly.

I do not suppose that personal advancement is a principal motive for cheating in science: rather it is the hunger for scientific reputation and the esteem of colleagues. And I believe that the most important incentive to scientific fraud is a passionate belief in the truth and significance of a theory or hypothesis which is disregarded or frankly not believed by the majority of scientists—colleagues who must accordingly be shocked into recognition of what the offending scientist believes to be a self-evident truth.

* W. Broad and N. Wade, *Betrayers of Truth: Fraud and Deceit in the Halls of Science* (Century, 1983).

Two scientific theories or viewpoints are notorious for arousing this passion: the doctrine of the inheritance of acquired characters associated with the name of J.-B. P. A. de Monet, le Chevalier de Lamarck, on the one hand, and the farrago of sillinesses that may be compendiously called 'the IQ nonsense', on the other. Consider Lamarckism first. One kind of Lamarckian inheritance is so commonplace and obvious as to be recognized for what it is by anyone who gives the matter a thought: it is that in which parents or members of a parental generation impart to their children or in general to a filial generation the knowledge and skills they had themselves acquired during their lifetimes. This is heredity all right, but it is 'exogenetic' in character, in the sense that it is not mediated through the genetic plant of chromosomes and genes, but through precept, example, and deliberate indoctrination. Unlike ordinary or endogenetic heredity, this other kind is reversible and is Lamarckian in style, for that which is acquired in one generation may be transmitted to the next and so on, cumulatively. The existence of this mode of heredity has given people a powerful incentive to believe that ordinary or genetic heredity works in this way too, as it seems only natural justice that it should, and even professional biologists have been taken in by the parallel between exo- and endo-genetic heredity and by what looks like a constitutional inability to realize this is not how nature works. The mechanism of heredity is selective, not instructive: what happens in an organism's lifetime, even if it is a profound bodily modification brought about by an adaptive response, cannot be imprinted upon the genome. There is no known or even conceivable genetic process by which DNA can be taught anything. It seems most unjust that this should be so, but so it is, for in heredity a person's exertions to improve his body or mind to adapt himself to new environments all go for nothing.

Lamarckian inheritance is a topic upon which literary people have for some reason felt themselves entitled to express an opinion. It is entirely understandable that George Bernard Shaw should have done so, but less obvious why Samuel Butler should have been among their number, especially as he expressed better than anyone else the essence of the teaching of that August Weismann who overthrew Lamarckism. Butler said that according to Weismann 'a hen is simply an egg's way of making another egg'. Lamarckism has on at least one occasion been the subject of a scientific fraud, as recounted in Arthur Koestler's *Midwife Toad*. Koestler disclaimed being a Lamarckist, but created an atmosphere favourable to Lamarckism by representing Darwinian geneticists as the spokesmen for a dull and unperceptive establishment of conventional belief. Biologists and psychologists who have been won over to Lamarckian thinking have published a whole number of experiments which purport to demonstrate Lamarckian inheritance, but all such demonstrations have been faulty. Either they have not excluded an orthodox Darwinian interpretation, or they have been open to explanations of other kinds, or they have been technically faulty. I myself was involved as a spectator in one such attempt. There was no dishonesty, indeed nothing more culpable than self-deception—in this case, the enthusiastic selection, from results that were all over the place, of only those that fitted the hypothesis the experimenter was seeking to corroborate.

Why should Lamarckism arouse such passionate conviction of its truth? I believe that the well-known association of Lamarckism with the sinister and indeed evil opinions of Trofim Denisovich Lysenko points to a political explanation. Lamarckism seems only fair: is it not right that mankind should benefit from their exertions and utterly wrong that man's genetic provenance—his breeding, in fact—should determine absolutely his character, capabilities, and deserts? It was his well-founded suspicion that his teachings tended to question the pre-eminence of a man's breeding that caused Napoleon's contemptuously dismissive attitude towards Lamarck. To a man convinced that Lamarckian inheritance is true because it is fair and socially just, it seems that selectionist theory presents an attempt, in Condorcet's words, to 'render nature herself an accomplice in the crime of political inequality'.

The second of the two major causes of that passionate belief that can conduce to fabrication was that which I referred to as 'the IQ nonsense', by which I mean the complex of beliefs arising out of a contention of H. J. Eysenck's which I lose no opportunity to hold up to public ridicule. 'Clearly the whole course of development of a

child's intellectual capabilities is largely laid down genetically . . .'
(*The Inequality of Man* (1973), p. 111).

The most shocking deception arising out of a passionate acceptance
of the idea that intelligence is susceptible to a scalar measurement and
is 90 per cent heritable was the lengthy and studied scientific frauds
of Sir Cyril Burt in his measurement of the IQs of twins reared
together or separated from birth. This fraud was uncovered by
Dr Leon Kamin and a skilful geneticist turned investigative journalist,
Dr Oliver J. Gillie. In the present book, which gives a lot of attention
to this case, Broad and Wade illustrate the inefficacy of scientific
monitoring within the profession itself—of the procedures which
those of us who maintain the integrity of scientists believe prevent or
rectify scientific fraud. But the reason Burt's findings were not
subjected to intent and independent critical scrutiny is simple and
understandable. There was no effective check of Burt's findings
because he told the IQ boys exactly what they wanted to hear. The
fault lay not with the scientific monitoring system but with the
bigotry and deep-seated misconceptions of the champions of the IQ
concept.

The present authors greatly enlarge our understanding of the
Burt frauds by recounting how a graduate student of Iowa State
University, Leroy Wolins, 'wrote to 37 authors of papers published
in psychology journals asking for the raw data on which the papers
were based'. No fewer than 28 reported that their data had been
misplaced, lost, or inadvertently destroyed.

The difficulty of laying hands on the 28 sets of data that were 'lost' or
withheld was made somewhat more comprehensible by the horrors that
emerged from the nine sets made available. Of the seven that arrived in
time to be analysed, three contained 'gross errors' in their statistics. The
implications of the Wolins study are almost too awesome to digest. Fewer
than one in four scientists were willing to provide their raw data on request,
without self-serving conditions, and nearly half of the studies analysed had
gross errors in their statistics alone. This is not the behaviour of a rational,
self-correcting, self-policing community of scholars.

Burt's is only one of the many notorious cases of fraud the authors
deal with. All the old favourites are to be found in the index:

Piltdown, Paul Kammerer of *The Midwife Toad*, and the infamous William T. Summerlin.*

A shocking story. Yet it is not the authors' intention to shock, though in fact they do so; no, the purpose is rather to show that research is not a wholly rational and explicitly logical procedure but subject to the confinement and constraints that afflict other professional men trying to make their way in the world. Moreover it questions our comfortable assumption that scientific cheating is very rare—an exceptional event that does not become a serious threat because science is protected by a whole number of built-in professional safeguards which bring it about that fraud is soon uncovered and the culprit punished.

The authors are very experienced professional science writers, and have made a highly responsible and well-argued contribution to the sociology of science. In spite of these sterling virtues, their book contrives also to be interesting and readable, and suitable for a lay readership. Even science writers are sometimes frauds, and the present authors, though they must have been aware of it, make no mention of a book by a science writer giving an account of a professedly authentic example of human cloning, ingeniously tricked out with quasi-scientific references put in 'to add corroborative detail to an otherwise bald and unconvincing narrative'.

What lesson should the scientific profession learn? Should we henceforth go around on our guard, doubting and questioning, looking for fraud and misrepresentation with the air of men expecting to find evidence of it? No, indeed not. Listening for a second time to Sir Kenneth Clark's splendid series of television broadcasts on 'Civilisation', I was again struck by the importance that Clark attached to confidence as a bonding agent in the advance of civilization, as it is indeed throughout professional life. Do not lawyers, bankers, clergymen, librarians, and editors tend to believe their fellow professionals unless they have a very good reason to do otherwise? Scientists are the same. The critical scrutiny of all scientific findings—perhaps especially one's own—is an unqualified desideratum of

---

* The Summerlin case is omitted here because there is a much fuller account in Chapter 8.

scientific progress. Without it science would surely founder—though not more rapidly, perhaps, than it would if the great collaborative expertise of science were to be subjected to an atmosphere of wary and suspicious disbelief.

8

The Strange Case of the Spotted Mice
(1976)

'Can the leopard change his spots?' the prophet asked (Jeremiah, 13:23), clearly not expecting to be told he can. Nor, indeed, can mice, except under the rather discreditable circumstances now to be outlined.

It is a well-attested truth of observation that except under special and unusual circumstances skin from one mouse or human being will not form a permanent graft after transplantation to another mouse or another human being; for although such a graft heals into place it soon becomes inflamed and ulcerated, and eventually dries up and sloughs off. The exceptional circumstances are: in human beings, when donor and recipient are identical twins, and in mice when prolonged inbreeding (e.g., upward of twenty successive generations of brother/sister mating) has made the mice so closely similar to each other genetically that they almost could be identical twins.

This being so, great surprise was caused in the world of transplantation when Dr William Summerlin, a member of the largest and in many ways the most important cancer research centre in the world, the Sloan-Kettering Institute in New York, with the backing of his chief, Dr Robert A. Good, made known in 1973 his surprising claim that a comparatively simple procedure—'tissue-culture'—could make a skin graft or a corneal graft from a member of the same or even of a different species acceptable to an organism that would otherwise have rejected it. This claim was specially important because grafting skin from one human being to another has never entered clinical practice, in spite of encouraging successes with the transplantation of kidneys, livers, and sometimes even hearts. Either skin is specially well able to excite the immunological reaction that leads to its own rejection, or it is specially vulnerable to it. This inability to graft new skin from one person to another is the greatest current shortcoming of the surgery of replacement and repair, because the replacement of skin is the only adequate treatment of extensive burns or excoriating wounds.

Summerlin's treatment, the technical details of which, in spite of exhortation from his director, he seemed suspiciously reluctant to impart to his colleagues, amounted in principle to very little more than the incubation of the intended graft in a suitable nutrient medium outside the body for a matter of days or weeks. This seemed an astonishingly simple solution of a problem no one else had solved, although many of us had been trying since about 1940.

Unfortunately, experienced biologists in other laboratories, and eventually workers in the same institute, were unable to confirm Summerlin's findings, so that Summerlin eventually had recourse to faking his results to convince his now uneasy chief. He touched up his grafts with a felt pen, so simulating dark skin grafts on white mice. He also claimed that operations had been done which had not been done. The formal end of the story came in 1974 when Dr John L. Ninnemann and Dr Good published a paper that in effect demolished the whole story.

This was all a nine days' wonder in the world of immunology, but the nine days are now up and this is therefore a good moment, on the basis of Hixson's very readable and, so far as I can tell, very accurate account* of the whole story, to stand back and see what lessons can be learned from the whole episode. This is also Hixson's ambition, for he says at the outset: 'I hope that by the time the reader has reached the end of the book, he or she will have enough information to form an opinion about what is good and what is not so good in our current system of medical research as it pertains to cancer.' 'If the reader disagrees with the author,' he goes on bravely, 'why then, so much the better.'

Summerlin's sin is not now in doubt; but it is still worth considering precisely why his action was considered so heinous by all his fellow

* Joseph Hixson, *The Patchwork Mouse* (Anchor Press, 1976).

scientists. The reason is this: scientists try to make sense of the world by devising hypotheses, i.e., draft explanations of what the world is like; they then examine these explanations as critically as they know how to, with the result that either they gain confidence in their beliefs or they modify or abandon them.

In the ordinary course of events scientists very often guess wrong, take a wrong view, or devise hypotheses that later turn out to be untenable. This is an ordinary part of human fallibility and calls for no special comment. Nor does it necessarily impede the growth of science because where they themselves guess wrong, others may yet guess right. But they won't guess right if the factual evidence that led to formulating the hypothesis and testing its correspondence with reality is not literally true. For this reason, any kind of falsification or fiddling with professedly factual results is rightly regarded as an unforgivable professional crime.

In trying unsuccessfully to get the same results as Summerlin, my colleagues and I wasted a lot of time that might have been much more fruitfully employed. Our failure—and the failure of others— to repeat his results was not in itself irremediably damaging, for this, too, is an ordinary part of scientific life. After Rupert Billingham, Leslie Brent, and I published experiments showing quite clearly that the problem of how to overcome the incompatibility barrier between unrelated individuals was indeed soluble, several people tried to repeat our work and failed. There were, however, always good reasons why they did so; either they had introduced into our techniques little 'improvements' of their own, or they were too clumsy or something. These failures did not disturb us in the very least: we knew we were right—and we were—so we did our best to tell those who were struggling with our techniques how best to carry them out. As Hixson makes plain, Summerlin was suspiciously at fault, for he simply would not divulge his methods. Indeed, matters reached such a point that Leslie Brent, one of the world's foremost experts on transplantation, was driven in desperation to send a whole file of his correspondence with Summerlin to Dr Good, an action unwillingly taken which led to Summerlin's being severely reprimanded.

A particularly exasperating characteristic of the whole episode was that Summerlin's claim *could* easily have been true and for reasons which Dr Good described as 'trivial'. They would have been trivial only because they did not point to any scientifically exciting phenomenon such as change of genetically programmed characteristics in the graft, for example its makeup of immunity-provoking substances. But from the point of view of clinical useful-ness it obviously didn't matter whether the explanation was profound or trivial. So many of us—even those who like myself shared Good's view that the reasons for the grafts' anomalous 'take' were trivial—persevered in trying to repeat Summerlin's work.

I am desperately sorry that Summerlin's work turned out to be mistaken because its failure means that we are still without means of repairing the skin surface except by piecemeal patching with little fragments of the patient's own skin—a process that may take weeks or even months during which the patient steadily loses body fluids and is specially vulnerable to infection.

The reader may well want to know what the very distinguished members of the Board of Scientific Consultants of the Sloan-Kettering Institute were up to all this time. My name appears repeatedly in Hixson's book as an expert on transplantation and partly as a member of the board. I cut a better figure in the pages of Hixson's book than I did in real life—something for which I bear Hixson no ill will. My reason for saying so is that at several critical points I found myself lacking in moral courage.

Summerlin once demonstrated to our assembled board a rabbit which, he said, had received from a human being a 'limbus to limbus' corneal graft—a graft which had been made compatible by his process of culturing. 'Limbus to limbus' means extending over the whole dome of the cornea to the extreme rim in which the blood vessels run. Through a perfectly transparent eye this rabbit looked at the board with the candid and unwavering gaze of which only a rabbit with an absolutely clear conscience is capable.

I could not believe that this rabbit had received a graft of any kind, not so much because of the perfect transparency of the cornea as because the pattern of blood vessels in the rim around the cornea was in no way disturbed. Nevertheless, I simply lacked the moral courage to say at the time that I thought we were the victims of a hoax or confidence trick. It is easy in theory to say these things, but in practice very senior scientists do not like trampling on their juniors in public. Besides, it was still possible that for some reason, 'trivial' or otherwise, the story was true. However we made no secret of our inability to repeat some of Summerlin's experiments, so far as we were able on the basis of the very inadequate information we had.

On the one occasion when I visited Summerlin in his laboratory with his immediate coworkers and technical helpers, and asked a number of hostile questions, I noticed with some surprise that our duologue was causing the others quite a lot of amusement. In retrospect, and after learning from Hixson the part Summerlin's technicians and immediate coworkers played in showing up the counterfeit, I can now see that they were sardonically amused at Summerlin's being interrogated in this way. But the equally plausible hypothesis I formed at the time was that, being only human, they were in reality amused at the obvious discomfiture of an eminent visiting scientist who, from the nature of his position on the board, was 'one of *them*' rather than 'one of *us*'. But for whatever reason, I did not forthrightly express any grave doubts about the probity of the whole enterprise.

In cases such as Summerlin's it is the usual thing to go over the culprit's career to find premonitions in his early life of how he behaved later. Hixson has done a good job here, reporting upon an unproved charge that Summerlin cheated in exams during his sophomore year at Emory University School of Medicine. It is, of course possible that Summerlin was what is known in the world of criminology as a 'bad apple', but this diagnosis lacks psychological depth.

I believe that there is a fairly simple explanation of Summerlin's egregious folly. It is this: in his early experiments Summerlin *did* actually obtain with mice, the results that later aroused so many misgivings. Mice can sometimes get muddled up even in the best regulated laboratories, and it is just conceivable that in his earliest experiments the recipient mice which Summerlin believed to be genuinely incompatible with their donors were in reality hybrids between the strain of the graft donor and some other mouse.

For genetic reasons, such hybrids would have accepted the donor's skin grafts anyway—irrespective of the 'tissue culture'; if this is what happened then Summerlin would naturally have been distraught when, on repeating his experiments with genuinely incompatible mice, he found they didn't work. Being absolutely convinced in his own mind that he was telling a true story, he thereupon resorted, disastrously, to deception. A mistake exactly analogous to this was once made by a securely established American expert on transplantation who reported an equally implausible result to the Transplantation Society; so far from losing face—except perhaps in his own mirror—this young man gained some credit for withdrawing his results and franky admitting that he had made a booboo.

Every scientist is at all times aware of his own fallibility and of the special safeguards that must be taken to avoid biasing the interpretation of results in a way that favours some hypothesis he may be temporarily in love with. 'Leaning over backward' is a well-known formula for avoiding self-deception—it stands for making sure that errors of observation arising from uncontrollable sources always tell *against* the hypothesis we should like to see corroborated. It is for this reason, too, that clinical trials of new remedies have to be done 'double blind'. Neither the patients nor the clinicians must know which patients are receiving a new wonder drug and which a mere placebo. A disinterested third party holds the key and will not unlock the code until clinical assessments are complete, after which it may, unhappily, turn out that 250 mg per day of placebic acid is as good a preventative of common colds as ascorbic acid (vitamin C).

My interpretation of Summerlin's behaviour is not intended to depreciate the importance and incentive to a scientist of public recognition and the esteem of one's colleagues. Election into the Fellowship of the Royal Society or its equivalent elsewhere is an honour because it is a public recognition of the fact that one's colleagues admire and applaud one's work. What is still mysterious is that a man of Summerlin's obvious intelligence and ability could have supposed he would get away with it, unless indeed, as I have suggested, he *did* once get the results he ultimately faked and thus felt perfely confident that workers elsewhere would, in the fullness of time, uphold his claims.

How does Robert Good, the director of Summerlin's institute, come out of all this?

There is no more important position in medical science than the directorship of the Sloan-Kettering Institute, a position for which Dr Robert Good was qualified by being enormously knowledgeable, brilliantly clever, persuasively eloquent, and indefatigably hard-working. Such a man naturally accumulates enemies, many of whom must have felt several inches taller when the Summerlin affair threw discredit upon him. Quite the most sickening aspect of the whole business was the way in which so many people whose lives had, until then, been devoted single-mindedly to self-advancement sprang into moral postures, pursed their lips, and moralized in a vein of excruciating triteness and dullness.

The Summerlin affair, we were told, was the natural consequence of the prosecution of science in an abrasively competitive world with limited research funds. If Good attached such importance to the work and sponsored his colleague so eloquently in public, then why did he not take more pains to supervise the research and make sure that everything was as it professed to be?

The answers to these questions do not wholly exculpate Good (and he himself does not think they do), but they show him in a very much better light than his enemies would like. In the first place it is a very much more endearing trait in the head of an institute to champion and promote the interests of his young than to let them get on with it while he busies himself with his own affairs. It was, indeed, Good's patronage in Minneapolis that made it possible for Summerlin to have a career at all. In the second place it is not physically possible for the head of an institute of several hundred members to supervise intently the work of each one. Most such directors assume—and as a rule rightly—that young recruits from good schools and with good references will abide by the accepted and well-understood rules of professional behaviour. I write here with the authority of someone who has been the head of a large research institute, and who has himself been once or twice deceived by an impostor.

Summerlin's attitude toward Good is made clear by his statement that he went to the Sloan-Kettering because Good was there: 'In retrospect I have to plead guilty to an overdose of hero worship. You know, I felt very close to this man. Regrettably, it wasn't mutual, as it turned out.' If this last, mean-minded comment is characteristic of Summerlin, it makes much else intelligible.

An especially attractive feature of Hixson's book is the evidence of the way in which science writers collude with each other and with scientists to create an atmosphere that will help them raise funds for their research. Summerlin was evidently a beneficiary of this process and it shows great forbearance on Hixson's part that he is not more indignant than he is on behalf of his fraternity of science writers.

Hixson's writing is of the quality we have come rather complacently to expect nowadays from first-rate professional science writers, although there are occasional lapses. On one page Summerlin is described as a 'tall, balding young skin specialist', in the idiom *Time* magazine has accustomed us to. Would it have mattered if Summerlin had been stocky and bushy-haired? Perhaps, for elsewhere Hixson makes it pretty clear that Summerlin's charm, enthusiasm, and general plausibility were not the least important part of his overall strategy for self-advancement.

The help science writers can give is very important since nearly all biomedical research—and particularly cancer research—is enormously costly, and the passages in which Hixson describes how it is financed will particularly interest not only potential private benefactors but also those needy scholars, especially in the humane arts, who follow less richly endowed pursuits. By one means or another—whether by private benefactions or fiscal levies—it is the general public that finances all cancer research, and this is how it should be, for they are ultimately the beneficiaries.

By the standards prevailing in humanistic studies the sums available are very large, but they are not to be had merely for the asking: one or more—sometimes a tier—of expert committees, all notoriously hard to please, stand between the applicant and the moneybags. Administrative committees refer a grant application to the expert body most highly qualified to express an informed opinion upon it and sometimes to other experts interested in cognate research.

Service on the National Institutes of Health Study Sections in America and on various boards of the Research Councils in England is enormously laborious and time-consuming, particularly as it may sometimes, as with Summerlin's application, entail personal visits to the laboratories in which the applicants are working. But so well understood is the importance of the task these study boards perform that it is possible to recruit to them extremely busy and able young scientists, often at stages in their careers when they can ill afford to give up their time for the purpose. Among these young bloodhounds are often a number of seniors who, having themselves probably been beneficiaries, have seen it all before and can therefore help to prevent hasty or injudicious decisions.

The study group that visited Summerlin's laboratories evidently had some misgivings about the authenticity of the work, but however deep-seated these may have been Summerlin was funded both by the NIH and by other benefactors.* Does this point to something deeply wrong with the present system of research funding—to something which, for the protection of the public, should be remedied right away?

Speaking as a man who has been a beneficiary of NIH research funds and who has served many severe sentences on grant-giving bodies, I should say not. In cost benefit, I should say that the most successful grant-giving bodies and research sponsors in the world are the Max Planck Gesellschaft in Germany and the Medical and Agricultural Research Councils in Great Britain. They have made grievous blunders, of course—including blunders of the kind that grant-giving bodies most greatly dread, i.e., the failure to fund research that has ultimately turned out to be of the greatest importance. But in the main they do whatever can be expected of bodies of highly informed and concerned scientists. Fortunately, clairvoyance and mind-reading are beyond them. It is not logically possible to predict future theories or future ideas, or, therefore, to be sure that a particular theory proposed for investigation will yield a harvest of fruitful ideas that will stand up to determined self-criticism.

* Dr Summerlin was not in fact funded by the National Institutes of Health nor other benefactors, but from Dr Good's discretionary funds—a fact pointed out by Mr Hixson and Dr Robert S. Schwartz in the *New York Review*, 10 June 1976, and accepted by Sir Peter.

However, there is an ignorance—amounting sometimes almost to contempt—of scientific philosophy not only among scientists but also among people, professedly critical thinkers, who ought to know, and often profess to know, better. This has led to the widespread misconception that the scientist works according to the rules of some cut and dired intellectual formulary known as 'the scientific method'. It has therefore come to be widely believed that given money and resources a scientist can bend the scientific method to the solution of almost any problem that confronts him. If he does not, it can only be because he is lazy or incompetent. In real life it is not like that at all. It cannot be too widely understood that there is no such thing as a 'calculus of scientific discovery'. The generative act in scientific discovery is a creative act of mind—a process as mysterious and unpredictable in a scientific context as it is in any other exercise of creativity.

We cannot devise hypotheses to order. Shelley would have understood this perfectly, for in his *Defence of Poetry* he wrote: 'A man cannot say "I *will* compose poetry"; The greatest poet even cannot say it. . . .' Nor can even the greatest scientist undertake to have illuminating ideas upon any problem he is confronted with, though he will know probably from experience how to put himself in the right frame of mind for getting ideas, and what reading and discussions will help him have them.

For these reasons most grant-giving bodies have come empirically to understand that they are most likely to do good by supporting people rather than projects, though I myself think this is a confession of weakness, for while conceding that no committee can do research, I nevertheless think that a committee which really knows its business should be able not merely to formulate a problem but also to indicate the lines along which it is most likely to be solved. Even if it were wrong, as very likely it would be, its thinking on the matter might easily spark off some fruitful idea in the mind of the investigator it commissioned to undertake its project.

I do not, however, think there is anything radically wrong with our present grant-giving procedures or that the grant-giving agencies can be convicted of anything more serious than of sometimes making mistakes. The romantic view of the creative process of science as something cognate with poetic invention is often sneered at by people who pride themselves as shrewd, practical-minded men of the world with a sound sense of the value of money. But they don't do any better than the rest of us, and it is they, indeed—people who believe that there is a cut and dried scientific method and that they can buy scientific results by paying for them—who are the incurable daydreamers with their heads in the clouds and no real understanding of the way the mind works.

It is characteristic of Hixson's balanced and fair-minded account of the Summerlin affair, which is likely to be the definitive account of the whole business, that he cites criticisms of his own profession. His discussion of the relationship between scientists and the press during an annual meeting of the Federation of American Societies for Experimental Biology makes it obvious that science writers want clear stories without the cagey reservations scientists are always introducing, and this may do something to incite people of Summerlin's temperament—though goodness knows he needed little encouragement.

How do scientists come out of it all? When the Summerlin affair became known, laymen shook their heads regretfully and exchanged long, significant looks as if to imply that they had learned something profoundly new about the scientific life and the morals of scientists. This is because two stereotypes of scientists dominate the lay imagination: the first is a figure like Martin Arrowsmith with a chronically dedicated expression on his face who is willing to sacrifice wealth and an easy life, even love and personal advancement, to the discovery of the new serum upon which he is covertly working after his colleagues have left college or laboratory for the night. The second is a Gothic figure intent on devising ever more expeditious means of destroying the human race—a man who, as his work comes to fruition, cries out in a strong Central European accent (for no American or Britisher could be guilty of such behaviour), 'And soon ze whole vorld vill be in my power' (maniacal laughter). In reality there are all kinds of different people who are scientists. I once put the matter thus:[1]

---

[1] *The Art of the Soluble* (Methuen, 1967).

Scientists are people of very dissimilar temperaments doing very different things in very different ways. Among scientists are collectors, classifiers and compulsive tidiers-up; many are detectives by temperament and many are explorers. Some are artists and others are artisans. There are poet-scientists and philosopher-scientists, and even a few mystics.

If only I had thought to add '. . . and just a few odd crooks', then I should have drawn a clear distinction between the scientific profession and the pursuit of mercantile business, politics, or the law, professions of which the practitioners are inflexibly upright all the time. As it is, I am afraid no great truth about scientific behaviour is to be learned from the Summerlin affair except perhaps that it takes all sorts to make a world.

9

Creativity—Especially in Science
(1985)

Creativity is the faculty of mind or spirit that empowers us to bring into existence, ostensibly out of nothing, something of beauty, order or significance. It is because of the high romantic illusion, to be dispelled later, that creativity builds God-like upon nothing that the process has been held in awe and thought divine in character from the days of Plato onwards. Plato spoke somewhere of the 'divine rapture of creation' and somewhere else of the 'divine fury' of creativity. In that deliciously ironic dialogue the *Ion* Socrates teasingly induces the braggart Ion to acknowledge that he is divinely inspired—is indeed a mouthpiece of divinity. In the same spirit Coleridge declared the creative imagination to be 'a repetition in the finite mind of the eternal act of creation in the infinite I AM'. There is no clearer affirmation of the romantic conception of the creative process.

The 'illusion of nothingness' as I may call the notion that that which is created arises out of disorder, chaos or nothingness was dispelled once for all by the American literary scholar John Livingston Lowes in *The Road to Xanadu**—a book worth the sum total of everything else that has been written on creativity by anyone at any time, including Plato.

Lowes's great literary feat was to uncover the origins in Coleridge's thinking and readings of the marvellous imagery of his two most famous poems, *the Rime of the Ancient Mariner* and *Kubla Khan*. Lowes also traces for us the origin of certain pure fancies that cannot have been based on observation because they refer to phenomena which, being physically impossible, cannot have occurred:

* (New York, 1927)

> . . . till clombe above the eastern bar
> The horned moon with one bright star
> Within the nether tip.

This is pure fancy of course, a mere traveller's tale: it is astronomically impossible that a star should ever be framed between the horns of the moon. Coleridge himself would have gone along with this judgement upon his fancy for he himself says of fancy that it is no other than a mode of memory emancipated from the order of time and space. Lowes's great achievement was to have reinstated the coordinates of time and space in the creatures of Coleridge's fancy.

I do not suppose that anyone could propound a hypothesis of the nature of creativity more likely to arouse a derisive titter than the idea that in the act of creative imagination we are mouthpieces of divinity—how *could* anyone entertain a concept so far removed from science—a concept lying so far on the wrong side of the tracks created by Karl Popper's line of demarcation between the domains of, on the one hand, science and common sense and on the other hand, of religion, metaphysics, and imaginative literature. Let us have a good laugh—but only in the vivid consciousness that we certainly cannot do any better and that the explanations of creativity by psychologists and the machine intelligence boys are in their own way just as inadequate.

The question of the nature of creativity poses a real challenge: how should a scientist address it? It is a question he should try to answer for his own sake because a scientist's explanations and hypotheses are all creative exploits: a scientist also exercises a creative imagination which, in Shelley's view, to which I shall refer later, is cognate with a poet's.

Suppose for the sake of argument that a scientist were hired to explain the nature and origin of creativity, how would he go about it? His first thought, I suppose, would be to try to build up a phenomenology of creative thought to get an idea of what goes on in the process, never mind how. Does that which is created really arise out of nothingness—out of a *tabula rasa*?—is creativity indeed a writing on a clean slate? This was the question answered by John Livingston Lowes in a form that I believe to be universally valid although it related directly only to Coleridge's two most famous poems; and then he would want to ask himself whether a creation arises in a form close to, or rather far removed from, that which we later come to recognize as its definitive form.

As to methodology—certain scientific precepts and counsels that will be in the forefront of the mind if indeed they had ever been relegated elsewhere:

1. We must study particulars and not abstractions: after all biologists do not study the nature of life: they study living things and likewise physicians study not the concept of illness but sick people; I myself am not engaged in the unending struggle against ignorance and disease but in trying to find out exactly why some esters of vitamin A strongly discourage the growth of tumours. The scientific student of creativity would accordingly investigate a number of different kinds of creative episodes to see what he could find in common between them.

2. A scientist will shun an explanation which, while it has outwardly the form of an explanation, does no more in fact than interpret one unknown in terms of another. It is fruitless, for example, to say that creative activity is the consequence of an act of sublimation or—if the creative person is a man—that it arises out of his discomfiture at having been unable to slay his father and supplant him in the marital bed.

3. A scientist will take evidence from all quarters likely to be informative, not excluding introspection, for no good would come of self-righteously abjuring such an important source of evidence.

4. A scientist must be resolutely critical, seeking reasons to disbelieve hypotheses, perhaps especially those which he has thought of himself and thinks rather brilliant.

I begin now by taking evidence from myself and find that the first thing I can say for certain about creativity is this; whatever the nature of the process may be, it is very vulnerable to error—decisive evidence against a divine origin. I say this with confidence because so very many of the hypotheses I have propounded turned out to be mistaken. Persons temperamentally inclined towards the notion that inspiration is of divine origin would certainly not accept the entailed consequence that God sometimes makes mistakes. A second truth of introspection that has struck me very forcibly is that those hypotheses which have stood up to critical assaults and have, so to speak, been received into the repertory, came into my mind fairly suddenly and in a form not very different from that which they ultimately took. A further consideration is this: every one of the hypotheses that I came to accept had long been the subject of lengthy fretful speculation, a kind of turbulence of mind—nothing arose genuinely *de novo* in my mind and I think it hardly conceivable that it should have done so.

I wonder therefore if the wholeness or nearly-wholeness of the conception and the suddeness of its irruption into the mind are characteristic of creative activity in all its forms. Let us therefore turn to consider and annotate one or two creative episodes of other kinds.

What about creativity in science? I believe—though it would take a research as long and detailed as that of John Livingston Lowes to uphold it—that every variety of creative episode in the world of letters or of everyday life can find its equivalent in science. As a simple representative example of a creative process consider *wit* in its most familiar form, a sudden juxtaposition of seemingly incongruous ideas that may be on the one hand laughter-provoking or on the other hand deeply illuminating or informative.

The example of wit I propose to recount turns on the ambiguity of the word 'premise'. The word 'premise' may refer to the assumption underlying a logical argument or to that argument's starting-point as in the context 'I do not accept your argument because its premises are altogether faulty'. Or, of course, a 'premise' may refer to a building, property, or occupied land as in the notice displayed throughout the London Zoo: 'Persons found feeding the animals will be escorted immediately from the premises.' The Reverend Sidney Smith was one day walking through the narrow streets of old Edinburgh when a furious altercation was heard between housewives addressing each other high up across the street where the buildings lean toward each other. Sidney Smith and his friend listened for a while and then

Smith said 'They can never agree, sir, for they are arguing from different premises.'

How shall we match this in science? Not at all, I fear, for its laughter-provoking properties, because science is a grave matter that does not lend itself to belly laughter. It is therefore some other property we must look for, a property such as the power of wit to illuminate or to edify. My first example comes from the English zoologist Hans Kalmus. When I was a student a favourite topic of discussion was 'What is the nature of the gene?' 'Is it a particle of some kind, an enzyme, the primer of a metabolic process or what?' At that time it was of course impossible to say with strident self-confidence that a gene is a singularity in the linear ordering of nucleotides along a lengthy polymeric molecule compounded of phosphoric acid, a five carbon sugar, and an organic base. The answer we really wanted was one that would have speeded up genetics greatly if it had been known and its implications had really sunk in and was very nearly propounded by the great physicist Erwin Schroedinger but was in fact precisely articulated by Hans Kalmus in the following terms: *a gene is a message*. Thus in looking for particles, primers, or enzymes we were looking for entirely the wrong things for one can't dissolve messages as one can dissolve deoxyribonucleoproteins in distilled water and then precipitate them by adding sodium chloride to a concentration of 0.15 molar. It was Francis Crick and James Watson who translated Kalmus's flash of wit into physical terms by interpreting the crystalline structure of deoxyribonucleic acid and showing how that structure qualified it to convey a genetic message.

A second example of scientific wit comes not from a scientist but from an English novelist, Samuel Butler, writing at a time when the doctrine of the inheritance of acquired characteristics—'Lamarckism'—was still taken seriously. Butler expressed better than ever before the essence of the teaching of August Weismann on the continuity of the germ plasm and the inability of the soma to influence the germ directively: 'A chicken', said Samuel Butler, 'is only an egg's way of making another egg.' This is one of the most profound truths in the whole of biology and it has never been better put.

But it is not examples such as these that people think of when they refer to creativity. They think rather of poetic inventions, of that kind of poetic imagery which in Sir Philip Sidney's words, 'doth strike, pierce and possess a sight of the soul'. Accordingly I choose an example from Shakespeare that I think will be found to illustrate the wholeness and lack of premeditation that is in my experience characteristic of scientific invention.

In *Macbeth*, Act 2, scene 2 Macbeth has murdered Banquo (a bloody business) and his hands and arms are covered with blood. 'Go get some water,' says Lady Macbeth, 'and wash this filthy witness from thy hand.' It sounds easy enough: just go to a washbasin and hold the hands under a running tap until they are clean; but there was no running water in medieval Scottish castles, only washbasins with pitchers of water. Macbeth must have used many of these and reddened basin after basin with blood. He thinks bitterly to himself that even if he were to wash his hands in the sea that too would be made bloody. 'Will all great Neptune's oceans wash this blood clean from my hand?' he cries: 'No, this my hand would rather the multitudinous seas incarnadine, making the green one red.'

I do not think this was consciously premeditated, that is, worked out beforehand. The words 'multitudinous' and 'incarnadine' appear nowhere else in Shakespeare and there is no evidence that Shakespeare ever wrote a memoir on the plumbing deficiencies of medieval Scottish castles.

The phenomenology of creation is all very well but what is the mechanism of it? I am sorry to say that we have nothing to add to the conclusion drawn by John Livinston Lowes: 'the ways of creation are wrapped in mystery; we may only marvel and bow the head' . . . but let us hear two dissentient voices.

Herbert Simon, whom I have heard lecture on this subject, believes that scientific discoveries are essentially of the nature of problem-solving and thus lie in principle within the capability of a computer. He refers especially to Dr Pat Langley's *Bacon 3* program remarking that this inductive machine is programmed to discern correlations and invariants and can if fed with the right numerical data rediscover Boyle's gas law and Kepler's third law of planetary motion.

This is not creativity, though, because the process does not flout the law of conservation of information and no new element of order is brought into being. The order was implicit in the data fed into the machine and needed only to be brought out, much as the theorem of Pythagoras is implicit in the axioms and postulates of Euclid, needing only to be brought out into the open; it is not new, it is already there: there is no creative act. Besides, is it conceivable that—except by mistake—a computer should say something witty in the sense in which the epigrams of Hans Kalmus and Samuel Butler were witty? Consider therefore a mistake by a computer. A computer designer proudly demonstrating to a friend a computer that would translate to and from English and Chinese: the friend asked it to translate into Chinese the English idiomatic phrase 'Out of sight, out of mind'. There followed a rumbling and humming noise and the computer printed out the answer. 'How do I know that this is correct?'—'Please translate it back from Chinese into English'. The computer did so and the English version of 'Out of sight, out of mind' turned out to be 'Invisible idiot'.

Dr Donald Campbell has a somewhat different conception of scientific discovery. Reflecting upon the truth that evolution is the only creative process of nature and that it is mediated through natural selection of genetic variants, Campbell reasoned that in human creativity a cognate process must be at work: human creativity must be a rapid combination and recombination and reassortment of ideas, the memory retaining the more plausible juxtapositions rather as if the computer were programmed to produce jokes of the random kind I illustrated above, while a selective process would have sorted out those that were genuinely funny from others that were merely silly or meaningless. I suppose it is possible that such a process does go on below the surface of the mind and that it is an element in the subconscious processes which must be presumed to precede the irruption of a hypothesis or explanation into the mind.

I do not find this way of looking at things very illuminating even if there is an element of truth in it. For me, the element of mystery John Livingston Lowes referred to is with us still.

I believe that there is no qualitative difference between creativity in science and in its many other contexts. For Shelley the act of creation, *poiesis*, is fundamentally the same in all its contexts and this consideration is what led him to say in his famous *Defence of Poetry* that 'poetry comprehends all science'. He meant that the kind of creative process that generates on the one hand poetry as ordinarily understood is also that which operates in the context of science, generating laws and explanations and all else that we recognize as the furniture of scientific thought.

I go along with this and admire it as a corrective to the widespread belief that science is a matter of facts and calculations and—in Wordsworth's words—'analytic industry'. Indeed, I feel in no way diminished by Shelley's conception and am pleased and elevated by the thought that we scientists have also a place in Parnassus alongside musicians and poets in company with Apollo and the Muses and others who create.

10

The Philosophy of Karl Popper
(1977)

It is convenient, though (as I shall explain later) somewhat artificial, to consider the philosophy of Professor Sir Karl Popper, FRS, FBA, under two headings: on the one hand, the philosophy of history and of social science, and on the other hand the philosophy of natural science.

Popper's Philosophy of History and of Social Science

Quite early in the nineteenth century the recognition that science and technology were incomparably the most successful endeavours human beings had ever engaged upon, nourished the ambitious belief that the methods which had been so successful in science and technology could be applied to human beings and to society.

It was widely hoped and believed that it would be possible to recognize and propound laws of the historic process and of social transformation—in effect, laws of historical destiny—which could have the same predictive value as the laws of physics and chemistry. When Edmund Halley predicted in 1705 that his comet would return in 1758, he was doing no more than social scientists might be able to do if only they were able to formulate the laws of historical change.

For beliefs of this general kind, Popper coined the word 'historicism'. In his first publication in the English language—a series of papers in *Economica*[1] entitled 'The Poverty of Historicism'—Popper showed that historicism was without logical foundation—was, indeed, mistaken. Popper's position in the social sciences may

[1] *Economica*, NS 11/5, 42, 43 (1944); 12/46 (1945); 3rd edn. (London, Routledge and Kegan Paul, 1957).

therefore be said to be analogous to David Hume's in the natural sciences, for Hume had caused a similar upheaval of accepted conventional thought when in much the same quiet, unassuming, and lethally effective way he destroyed the foundations of empiricism by showing that the law of causality and the principle of the uniformity of nature were based not upon logical reasoning but upon faith—upon a mere habit of expectation.

A Historicist Argument

I shall now give a real-life example of a fallacious argument based on historicist principles—an example which illustrates very clearly the kind of reason why Popper, in 'The Poverty of Historicism', rejected the possibility of devising predictive laws of human destiny.

What can the laws of economic determinism tell us about placing and design of factories and the dwellings that tend to grow up around them? Popper cites the following typically Marxist argument.[2] The argument goes as follows:

The source of energy is coal: therefore factories and the dwelling places that go with them must be situated near coalfields, to reduce to a minimum the economic burden of transporting coal.

Coal yields power through the agency of *steam*, but as it is hardly feasible for each machine tool in a factory to be powered by its own steam engine the designer of the factory will so contrive it that a single steam engine drives a single overhead shaft or spindle, which is connected by flexible belts to the individual machine tools on the shop floor.

Thus not only will factories be ever more concentrated in areas near coalfields, but individual factories will get larger and larger so as to get the maximum economic benefit from the operation of a single steam plant.

Unfortunately for theories of economic determinism, the predictions embodied in these arguments have not been borne out, for Marx was unable to foresee, and it would indeed have been impossible in principle for *anybody* to foresee, the advent of that electrical power which has made it possible for large numbers of small, as opposed to small numbers of large, factories to spring

---

[2] Unfortunately neither Sir Karl Popper, Mr Shearmer, nor I can remember the passage in Popper's writings in which this argument is discussed in detail.

up all over the place, not just in the immediate neighbourhood of coalfields, because the economic argument relating to coal transport is now no longer valid. This is precisely what has happened in the light engineering works of the English Midlands, and of the State of Massachusetts.

This whole argument illustrates Popper's reasoning at the very beginning of 'The Poverty of Historicism': human history is, at least to some degree, influenced by scientific and technological ideas, and inasmuch as these ideas are intrinsically unpredictable, one cannot predict the future course of history.

This means we must reject the possibility of a theoretical history; that is to say, of a historical social science that would correspond to theoretical physics. There can be no scientific theory of historical development serving as a basis for historical prediction. (p. vi)

It is very much in keeping with Popper's philosophy of history that he is deeply opposed to revolutionary or 'holistic' remedies for whatever may be amiss with society.[3] After all, a society—if it exists at all as a society—must already have a certain organization and working capability which makes it most unlikely that its imperfections can be remedied by bringing the whole of it tumbling down through revolutionary action or—what comes to the same thing—that it can be remedied by some kind of comprehensive overall treatment that changes society's entire structure into that of an organism of some quite different kind. The formula that has come to be especially associated with the name of Karl Popper is that of 'piecemeal social engineering'. It is not by revolutionary or holistic action that society will be improved. A multitude of things go wrong in any real society, and the important task of the social reformer is to find out in which particular way things have gone amiss and then, if possible, to put them right in such a way that he can find out whether or not his remedies have the desired effect . . .

The piecemeal engineer knows, like Socrates, how little he knows. He knows that we can learn only from our mistakes. Accordingly, he will make his way step by step, carefully comparing the results expected with the

[3] Holists believe that a society is an organism with so high a degree of integration of its parts that no tinkering with them is possible.

results achieved, and be always on the lookout for unavoidable, unwanted consequences of any reform; and he will avoid undertaking reforms of a complexity and scope which will make it impossible for him to disentangle causes and effects and to know what he is really doing. (p. 67)

Good real-life examples of social or economic engineering that have been embarked upon without a sufficiently anxious awareness of the possibility of untoward consequences are the cavalier attitude towards inflation of Keynesian economics, and at a less grandiose level, the Rent Acts which, though intended to produce one effect, sometimes had exactly the opposite effect.

The Open Society and Its Enemies

Anybody who dismissed the arguments of 'The Poverty of Historicism' as too abstractly philosophic in character must have been strangely lacking in sensibility, because from the Spanish Civil War onwards Europe seemed to have become a test-tube for the interaction between the two great historicist doctrines which prevailed at the time, Fascism and Marxism.

According to Marxist theory, the direction of the flow of history is shaped by a struggle for supremacy between social classes—in particular between the proletariat and those who own the means of production. It was argued that this struggle would inevitably lead to a social revolution, and end in the victory of the proletariat, with the disappearance of a class stratification.

The gist of Fascism cannot be summarized in any such familiar form of words, but the high Fascism of Nazi apologists such as Alfred Rosenberg was a form of racial or genetic élitism, which positively avowed that the advancement of mankind and the motivation of historic change were the special responsibility of a single race in the interests of which all opposition could justifiably be trampled down.

There is no need today to draw special attention to the misery and moral diminishment of man for which the political realizations of the two great historicist doctrines have been responsible, but in the early days of the war, and for some time after, there *was* a most urgent need for a philosopher to undertake a philosophic defence of democracy. This formidable task was discharged by Karl Popper in the work he wrote during his wartime exile in New Zealand: *The Open Society and Its Enemies*. Although it is a work of great gravity and expository detail, it is also very dramatic in character: it fired and inspired some of the more thoughtful young people of the day, and, more important, equipped them to rebut the dangerous argument that an old-fashioned democracy could not combat the menace of totalitarian states without becoming in some respects totalitarian itself.

We know from Popper's brief autobiography, in which he comments on the deadly dullness of some of the lectures he endured as a secondary-school pupil in Vienna, that Popper fully appreciated the importance of an arresting or even shocking presentation.

The Open Society was dramatic and shocking because of its temperate but nevertheless lethally effective criticisms of Plato, Hegel, and Marx. Popper criticizes them not because they are so many sacred cows, but because 'great men make great mistakes': if civilization is to survive we must break the habit of deference to great men. 'The aims of civilisation', he says, 'are humaneness and reasonableness, equality and freedom.' These great men, though we had been taught to admire or even revere them, had made these ambitions harder to realize.

An open society, such as our own, is a society in which disagreement and dissent, so far from being prohibited, are used as agencies of social improvement, for it is by criticizing intended legislation before it becomes statutory that we may hope to discover its imperfections in time to prevent ourselves from making serious mistakes. In an open society people can flourish in all their rich and sometimes strange diversity of political opinions, ethnic origins, and religious beliefs. In a closed society—Popper sometimes calls it a 'tribal society'—we are confined by submission to political forces or tribal observances. Only an 'open society' sets free the critical powers of man.

The Open Society was confessedly Popper's 'war work', and to equip himself for it and especially for his reappraisal of Plato, he set himself to study Greek again.

It was not to be expected that Marxists would greet *The Open Society* with a clamour of enthusiasm, but Popper went out of his way to ensure that a Marxist should review his book. The first review submitted to the *British Journal for the Philosophy of Science* was so laudatory that Popper, having some editorial say in the matter, proposed that the book should be sent to J. D. Bernal for reviewing.

The review[4] is most disappointing, and not all that one could have hoped from one of his remarkable intelligence. He chides Popper for misusing the word 'historicist', evidently unaware that Popper invented it, and he coins a new word of his own. Popper is a 'philagnoist'—a lover of ignorance—a description about as plausible as to call him the principal *tenore robusto* of La Scala, Milan.

Bernal makes some criticisms which make it pretty clear that he had not read 'The Poverty of Historicism'. In one of the criticisms, which shows that he had not read *The Open Society* as attentively as a reviewer should, he says that Popper accuses Marx of Utopianism, whereas in reality Popper, intent upon more serious criticisms, describes this misconception of Marxian theory as 'vulgar Marxism', part of a popular misapprehension of Marxist teachings.

From time to time one still meets people who have not read *The Open Society and Its Enemies* and I now accordingly beseech them to do so. Its arguments are still relevant to the modern world, and will continue to be so for as long as democracy is under threat from any form of tribalism.

The Philosophy of Science

Just as Popper's philosophy of history and of the social sciences grew out of his dissatisfaction with the complacently mistaken doctrines of historicism, so his philosophy of science could be said to have grown out of his criticism of the prevailing and orthodox opinion about how scientists go about their work to enlarge human knowledge: the opinion that scientists work by a special process known as *induction*.

Inductivism is not one but a complex of different notions, which hang together and are founded on the principle that the scientist is essentially a man who observes nature clearly and intently and

---

[4] 'Has History a Meaning?', *Brit. J. Philos. Sci.* 6 (1956), 104–67.

without any preconception in mind. The record of an observation is an 'observation-statement.' Science therefore consists of a majestic pile of observation-statements, and all scientific laws are compounded of these observation-statements by a system of rules of discovery— as John Stuart Mill called them—or more generally (and to make it sound more sophisticated) by what we may call 'a calculus of discovery'. The high priest of induction, John Stuart Mill, explained it in terms which, though not intended to do so, undermine it as effectively as David Hume had undermined it in his *Treatise of Human Nature* a hundred years before.

Mill said of induction ('that great mental operation') that it was a process of inference which 'proceeds from the known to the unknown; . . . any process in which what seems the conclusion is no wider than the premises from which it is drawn does not fall within the meaning of the term.'

As Mill also believed that induction was a logically rigorous process, so he is, in effect, telling us that by some rigorous act of mind it is possible to increase the empirical content of the assumptions it starts from. I don't think Mill quite understood what an audacious, even outrageous statement he had made, for if we can enlarge empirical information by an act of mind we scientists could conduct our research in a supine position with the eyes half closed.

Much else is wrong with induction. Using the word 'induction' now to refer to any scheme of thought which purports to show that general statements can be compounded of particular observation-statements, induction has no explanation to offer of two very familiar elements in scientific thought and discovery: error and luck, for why should we ever fall into error[5] if we observe things correctly and operate a calculus such as Mill's according to the rules?

As to luck, which all scientists benefit from, how can we *know* when we are being lucky except in terms of the fulfilment of some specific prior expectation?

Induction, moreover, is prey to a number of paradoxes, the existence of which is a sure sign that something, somewhere, is logically amiss. The most amusing we may call 'the paradox of the

[5] But here see K. R. Popper 'On the Sources of Knowledge and of Ignorance' in *Conjectures and Refutations* (London, 1963).

old black boot': philosophers are not very imaginative when asked for examples of an inductive generalization, and they are very apt to say: 'All swans are white', so let us proceed from there. If all swans are white, then all non-white objects are non-swans. (That is a cast iron logical inference.) This logical prediction is borne out by the discovery of an old black boot: it is not a swan, and it is not white, so our confidence in the generaliztion that 'all swans are white' is appreciably strengthened.

Popper's Solution of the Problem of Induction

Immanuel Kant, the foremost philosopher of *his* day, thought to remedy the shortcomings of empiricism by the attempted discovery and formulation of a priori knowledge—that is to say, of knowledge independent of all experience, and this great ambition is the programme of his *Critique of Pure Reason.* Popper's solution is to abandon altogether the search for apodeictic certainty—for truth that is demonstrably certain and therefore beyond criticism and he proposes instead that all scientific knowledge is conjectural in character.

In Popper's view the generative act in scientific discovery or in the solution of a problem is the formulation of a hypothesis, i.e. an imaginative conjecture about what the truth of the matter might be. A hypothesis is a sort of draft law or guess about what the world—or some particularly interesting part of it—may be like, or in a wider sense it may be a mechanical invention which we can think of as a solid or embodied hypothesis.

In the outcome science is not a collection of facts or of unquestionable generalizations, but a logically connected network of hypotheses which represent our current opinion about what the real world is like.

Most of the day-to-day business of science consists not of hunting for facts as an inductivist might suppose, but of testing hypotheses, that is, seeing if they stand up to the test of real life or—if inventions—to see whether or not they work. Acts undertaken to test a hypothesis are referred to as 'experiments'.

What is being tested in an experiment is the logical implications of the hypothesis, i.e. the logical consequences of accepting a hypothesis. A well-designed and technically successful experiment will yield results of two different kinds: the experimental results may square with the hypothesis, or they may be inconsistent with it.

If the results square with the implications of the hypothesis, then the scientist takes heart and begins to hope that he is thinking on the right lines; he will then, if he has any sense, expose his hypothesis to still more exacting experimental tests. The riskier the hypothesis, i.e. the more 'way out' or unlikely it seems to be in terms of current expectations, the more reassured the experimenter will be if it stands up to experiment; but no matter how often the hypothesis is confirmed—no matter how many apples fall downwards instead of upwards—the hypothesis embodying the Newtonian gravitational scheme cannot be said to have been *proved to be true*. Any hypothesis is still *sub judice* and may conceivably be supplanted by a different hypothesis later on. Many people find this element of Popper's philosophy rather disagreeable, and are inclined to think that some scientific generalizations are true beyond any serious possibility of further question.

Popper does not agree: in the whole history of science no theory has seemed more secure, more naturally and essentially right and less likely to be controverted than the body of theory that makes up Newton's celestial mechanics, but even this has been questioned and supplanted by the more general theory of Einstein. One of our foremost theoretical physicists, Professor Sir Herman Bondi, has said of Newton's theory: 'We may certainly speak of disproof now.'

The second possibility we may envisage is that the results of a carefully designed and well-executed experiment controvert the implications of the hypothesis. In real life this does not mean that the hypothesis is promptly abandoned; but some reconsideration is certainly called for which may, in the extreme case, amount to an abandonment of a hypothesis but is more likely to involve some modification of its current form. If, however, certain theories or observations are assumed to be true, then we may sometimes speak of the outright falsification of a hypothesis. The qualification 'if certain theories and observations are assumed to be true' is important, for falsification is not itself immune to error and something must always be assumed to be true if anything is to be shown to be false.

The Demarcation Problem

In one way falsification plays a specially important part in Popper's scientific philosophy. Suppose we put to ourselves the general question of what distinguishes statements which belong to the world of science and of common sense (we need make no distinction between the two) from metaphysical or fanciful statements?

Popper's answer would be that statements belonging to the world of discourse of science and common sense are in principle falsifiable and it must be possible in principle to envisage what steps we could take to test the statement and so maybe to find it wanting.

Popper does *not* go on to say (as in cognate circumstances logical positivists would have said) that the criterion of falsifiability-in-principle distinguishes scientific or commonsensical statements from metaphysics or nonsense. On the contrary: the line of demarcation is between the statements belonging to the world of science and common sense and statements belonging to some other world of discourse; so far from being nonsensical, metaphysical statements may lie on the pathway towards truth and may sometimes be conducive to the discovery of the truth.

To my mind the great strength of Karl Popper's conception of the scientific process is that it is realistic—it gives a pretty fair picture of what actually goes on in real-life laboratories.

In real laboratories there is no constant clamour of affirmation or denial. We are all very conscious of being engaged in an exploratory process as we cautiously grope our way forwards by the method which has come to be summed up by the now familiar cliché of *conjecture and refutation*.

Some people think the idea that science is conjectural in character in some way diminishes science and those who practise it; but to my mind nothing could be more diminishing than the idea that the scientist is a collector and classifier of facts, a man who cranks some well-oiled machine of discovery. Popper's conception of science is, in my opinion, a liberating one; I feel enlarged, not diminished, by the thought that any truth begins life as an imaginative preconception of what the truth might be, for it puts me on the same footing as all other people who use the imaginative faculty. I feel that what distinguishes the natural scientist from laymen is that we scientists have the most elaborate critical apparatus for testing ideas: we need not persist in error if we are determined not to do so.

The communications engineer would have no difficulty in recognizing Popper's formulation, the elements of a control process or a steering (cybernetic) process which accords exactly with the idea that the scientist is finding his way about the world. 'Feedback' is the fundamental strategem in all control systems, i.e. the control of a performance by the consequences of the act performed. If we regard the logical consequences we test by experiment as the logical 'output' of a hypothesis, then it is clear that the experiment which may cause a hypothesis to be modified or even in an extreme case abandoned, gives a textbook example of the phenomenon of negative feedback.

No scientist thinks of himself as a man of facts and calculations. Popper puts it thus: 'It is not his *possession* of knowledge that makes the man of science, but his persistent and relentlessly critical search for the truth.' I said right at the outset that though we can consider Popper's philosophy under two principal headings—the philosophy of history and of the social sciences, and the philosophy of the natural sciences—nevertheless I also said that the distinction was rather an artificial one. Why? What is it that they have in common?

What they have in common is the element that pervades the whole of Popper's philosophy: the recognition that human designs and human schemes of thought are very often (perhaps more often than not) mistaken, and that the safest way to proceed is to identify and learn from our mistakes and learn always to do better next time. By this means, Popper believes, as I also believe, the world can be made a better place to live in.

The Genetic Improvement of Man
(1972)

I make no apology for paying my respects fo Macfarlane Burnet in the form of a philosophic dissertation on the genetic improvement of man. Burnet is a virologist, an epidemiologist, an immunologist, and other things besides, but above all he is a biologist, a leader of biological thought. Because genetics, broadly conceived, is the central discipline of biology, and because evolution is its unifying doctrine, it follows that all biologists must be (or should be) interested in the notion of the genetic improvement of man. So having given some thought to the matter, I decided that no other theme could be more appropriate to the occasion.

It would be the merest naïvety to suppose that the idea of improvement—in its extreme or terminal form, of perfectibility—is a new one, or one that science has now authorized us to contemplate for the first time. The idea of improvement must be pretty well coeval with human speculative thought. In one form or another it embodies almost the whole spiritual history of mankind. No one man is qualified to give a complete account of the idea of improvement, even in the secular aspects to which I shall confine myself; but amidst all the profusion and confusion of thought on the subject it is, I think, possible to discern three main kinds of conception or vision about what the future of man in the world might be. I shall call them Olympian, Arcadian, and Utopian.

In the Olympian conception, men can become like gods; can achieve complete virtue, understanding, and peace of mind, but through spiritual insight, not by mastery of the physical world. No particular environment is envisaged as a setting for this apotheosis, and the environment itself need not have been perfected; the Olympian formula is for all seasons. In the Olympian vision, the direction of the human gaze is upwards or perhaps inwards; but in Arcadian thought, closely bound up with the ancient legend of a Golden Age, it is directed backwards. In Arcadia men remain human but in a state of natural innocence. They retreat into a tranquil pastoral world where peace of mind is not threatened, intellectual aspiration is not called for, and virtue is not at risk. An Arcadian society is anarchic; everything that is implied by authority is replaced by everything that is implied by fraternity. It is a world without strife, without ambition, and without material accomplishment.

From the seventeenth century onwards a new vision began to be taken seriously, the Utopian. Man can create anew and therefore improve the world he lives in through his own exertions; he begins as a tenant or lodger in the world, but ends up as its landlord; and as his environment improves, so, it is alleged, will he. Virtue can be learned and will eventually become second nature, understanding can be aspired to, but complete peace of mind can never be achieved because there will always be something more to do. Men look forwards, never backwards, and seldom upwards.

All three visions have both noble and comic elements, and each has developed its own satirical literature, which is often better known than the work it satirizes. In spite of these and other discouragements, scientists are characteristically Utopian in their outlook, because it is the only scheme of belief that makes sense of what scientists do. There is nevertheless one long-recognized weakness in Utopian speculation: the inadequacy of man, the extreme unlikelihood that man can live up to his own ambitions. It is for this reason that the idea of a genetic improvement of man has a special fascination for Utopian thinkers. One of the three princes of Campanella's *The City of the Sun*, a Utopia of the early seventeenth century, is named Love, and Love's business is to supervise a system of eugenic mating. 'He sees that men and women are so joined together that they bring forth the best offspring', Campanella's narrator says: 'Indeed they laugh at us who exhibit a studious care for our breed of horses and dogs, but neglect the breeding of human beings.'

The idea of genetic improvement as a realizable policy may be said to have begun with the writings of Francis Galton, the great nineteenth-century humanist who founded the science of eugenics, and coined the word itself. 'Eugenics', said Galton, 'is the science which deals with all the influences that improve the inborn qualities of a race; also with those that develop them to the utmost advantage.' 'Man is gifted with pity and other kindly feelings; he has also the power of preventing many kinds of suffering. I conceive it to fall well within his province to replace natural selection by other processes that are more merciful and not less effective. That is precisely the aim of eugenics.' We must understand that when Galton defined eugenics in these terms, he was combating a radical form of social Darwinism, especially championed by Ernst Haeckel, according to which the doctrine of the survival of the fittest applied to the social development of human beings no less than to the evolution of animal communities. A quotation from Haeckel himself will make my point: 'The theory of selection teaches us that in human life, exactly as in animal and plant life, at each place and time only the small privileged minority can continue to exist and flourish; the great mass must starve and more or less prematurely perish in misery. . . . We may deeply mourn this tragic fact, but we cannot deny or alter it.'

Social Darwinism in the form expounded by Haeckel provided a theoretical justification for the great biological crimes of Fascism, so it is hardly surprising that eugenics fell into complete discredit. Politically speaking its object was (I quote Condorcet) 'to render Nature herself an accomplice in the guilt of political inequality'. In spite of this, kindly and humane people, genuine descendants of Galton himself, have continued to believe that genetic policies can be used to improve the performance and capabilities of human beings, and that our power to use genetics for this purpose offers an antidote to the slow deterioration of man thought to be produced by the amelioration or softening of the environment, and is the foundation of all rational hope for the improvement of human society.

The case for 'positive eugenics', that is for constructive rather than merely remedial eugenics, is based on the model of stockbreeding. If horses, dogs and cattle can be improved by selective breeding, it is argued, why cannot human beings? (This is Campanella's question.) One can give two kinds of answer to this question—a moral and political answer, and a scientific answer. The moral-political answer is that no such regimen of genetic improvement could be practised within the framework of a society that respects the rights of individuals. This answer should be sufficient, but I am going to give the scientific answer also, for the following very important reason. Many people believe that it is scientifically feasible to create a population of supermen, and that only man's better self stands in the way of putting such a policy into practice. This belief, I hope to show, is quite erroneous, but it is important to refute it, because it is the kind of belief that underlies the modern conception of science as an essentially dehumanizing activity which may perpetrate some terrible mischief if it is not kept firmly under a control which it is at all times striving to elude.

In point of fact science has very little to do with it. The empirical arts of the stockbreeder are as old as civilization. Given a tyrant or a dynasty of tyrants, a scheme of selective inbreeding could have been enforced upon human beings at any time within the past five thousand years. It is not something that science has now for the first time put it within our power to do. At any time in the past few thousand years it would have been possible to embark on a scheme to make human beings as different one from another in appearance and capabilities as greyhounds and pekinese.

Let me now try to explain why a regimen of selective inbreeding is not scientifically acceptable, and why the stockbreeding analogy can no longer be sustained. In the old days (not so very long ago) the end-products of the stockbreeder's art—whether super-dogs or super-cattle or even super-mice—were expected to fulfil not one but *two* functions: two functions which, until recently, no one clearly distinguished and no one clearly realized could be separately and independently fulfilled. The first function was to be the end-product itself, to be the usable, eatable, or marketable goal of the breeding procedure. The second function was to be the parents of the next generation of super animals. In order to fulfil this second or reproductive function, the eugenic end-product had to meet a certain genetic specification, viz. that it should be homozygous or true-breeding in respect of all the characters for which selection had been exercised. If the animals were predominantly heterozygous or genetically mixed in their make-up, then the stock would not breed true and the efforts of the stockbreeder would be dissipated in a single generation.

Until recent years the dual ambition of the eugenic stockbreeder seemed to be upheld by genetic theory. The prevailing belief was that the animals belonging to a particular species or interbreeding community were predominantly homozygous in genetic composition. Natural selection was thought to be working towards the establishment or fixation of a particular genotype or genetic make-up, that which conferred the highest degree of adaptedness to the prevailing circumstances. If the circumstances changed, so also would the genetic make-up, because new genes—mutant genes—were continually proffered for selection, and these provided resources of variation rich enough to make it possible for natural selection to work out a new and improved genetic formula for survival. It is true that polymorphism was recognized as a departure from this tidy scheme (polymorphism refers to the stable coexistence of genetically differentiated types in the population for reasons other than the pressure of recurrent mutations); but polymorphism was thought of as a special phenomenon for the existence of which special explanations had to be devised. By and large, nevertheless, the stockbreeder and the theoretical geneticist supported each other's conceptions: genetic theory made sense of the stockbreeder's ambitions, and the stockbreeder could think of himself as a man who put genetic theory to practical use.

Today and for some years past this entire scheme of thinking has been called into question. It turns out that natural populations of outbreeding organisms, including human beings, are persistently and obstinately diverse in genetic make-up. Polymorphism is not an exceptional phenomenon but the rule. One drop of human blood can tell an astonishing story of human diversity. The haemoglobins and non-haemoglobin proteins, the red cell antigens, the serum enzymes and serum proteins generally, the leucocyte surface structures, all exist in a huge profusion of variant forms; and the same is almost certainly true of all the other macro-molecular constituents of the body. People have therefore come to abandon the view that natural selection works towards the fixation of a particular genotype, of some one preferred genetic formula for adaptedness. It is *populations* that evolve, not pedigrees; and the end-product of evolution, in so far as it can be said to have one, is itself a population, not a representative genetic type to which every individual will represent a more or less faithful approximation. The individual members of the population differ from one another, but the population itself has a stable genetic structure, i.e. a stable pattern of genetic inequality. Individual members of the population do not breed true, for being heterozygous their offspring will necessarily be unlike themselves. But, as G. H. Hardy pointed out more than sixty years ago, the Mendelian process is such that the population as a whole breeds true—i.e. reproduces a population of the same genetic structure as itself—even if its individual members do not. Under a regimen of random mating, the frequency with which the various genotypes appear in the population remains constant from generation to generation, except in so far as natural selection may change their proportions, and so cause a new pattern of genetic inequality to take shape. 'Natural selection' in this context refers simply to a state of affairs in which the different genetic categories do not make an equal contribution to the ancestry of future generations, i.e. a contribution proportional to their existing numbers. Some take a larger share and others necessarily a smaller share, and so the genetic make-up of the population changes.

This newer conception undermines the ambition of the old-fashioned stockbreeder, and makes nonsense of the eugenic ambitions that seemed to be supported by their practice. The stockbreeder is now seen to be undertaking an unnatural procedure, no longer authorized by our conception of the way that genetic changes happen in real life.

It is now known that this dilemma can be resolved, at least in principle. The stockbreeder must now no longer expect his animals to fulfil both the functions which, he supposed, went necessarily together: they cannot both represent the eugenic end-product *and* be the parents of succeeding generations.

To a layman the idea of dissociating the existential and reproductive functions of the eugenic end-product seems impossible to achieve except by some sort of conjuring trick—certainly impossible to reconcile with having eugenic end-products that are uniform in the characteristics they have been bred to possess. The trick is done, of course, by adopting a nicely calculated regimen of cross-breeding.

The principle is simple enough. Let me illustrate it from the world of laboratory animals. If I maintain two inbred strains of mice in my laboratory—strains sufficiently inbred to be homozygous and therefore true-breeding at most genetic loci—the animals representing the first generation of a cross between them will also form a uniform population. They will also probably be better performers than their parents in every way: in fertility, growth rate, intelligence, longevity, and resistance to disease. The characteristics of these first generation hybrids are specified by the genetic properties of the parental strains, and if the parents are in fact judiciously chosen, the hybrids may represent the eugenic end-product which we seek. They are, however, hybrid or heterozygous for every gene in respect of which the two parental strains differ. Therefore they cannot be bred from, because their own offspring will be not only diverse but maximally diverse, in the sense that they will exploit to the full the genetic possibilities defined by the make-up of the two parental strains. If therefore I wanted to raise a uniform population of super-mice at will, I should try to breed two homozygous strains whose F_1 hybrid progeny answered to my specifications, but they themselves, the progeny, would be relieved of a reproductive function, which would continue to be exercised by the two parental strains. Thus the dissociation between being the eugenic end-product and being the parents of the next generation would be complete.

Modern stockbreeding practice makes use of very much more sophisticated schemes of cross-breeding than this, but the principle is the same. The somatic and reproductive functions are separated: the eugenic end-product is reproducible at will, but does not reproduce.

This concludes my statement of the scientific reasons why the goal of positive eugenics, as Galton and his followers envisaged it, cannot be achieved. Human diversity is one of the facts of life, and the human genetic system does not lend itself to improvement by selective inbreeding. We could not adapt modern stockbreeding principles to a human society without abandoning a large part of what we understand by being human. Although negative or purely remedial eugenics has a useful and important function to fulfil in human society, I think that, in the main, for many centuries to come, we shall have to put up with human beings as they are at present constituted.

12

The Future of Man
(1959)

I. THE FALLIBILITY OF PREDICTION

The best way to give you an idea of the subject of these Reith Lectures is to put before you some of the questions I am hoping to answer, and this I shall do in a minute or two's time; but, first of all, for my own peace of mind, I should like to explain some of the uncertainties I have felt about their style and general purpose.

I first thought of attempting a grand prophetic statement about Man's future as *Homo sapiens*—of doing for biology something of the kind that physicists have done when they have written about the shape of the foreseeable future or of what might happen in the next million years; but I soon saw that if I were to attempt anything of the kind on behalf of biology, I should be obliged either to weary you with endless qualifications and reservations and disclaimers, or else to try to disguise the thinness of the reasoning by taking refuge in apocalyptic prose. The effect of the first procedure is to leave people perplexed about what the lecturer is actually saying, supposing he has screwed himself up to the point of saying anything at all; and of the second, to make them doubtful about his reasoning, stirring though what he said may well have been.

The more deeply I studied the problems I am going to talk about, the more deeply I became convinced that the opinions of the learned are often much less interesting than the reasoning which professes to uphold them; and in the outcome I decided that these lectures were to be about the *process of foretelling* rather than about what is actually foretold. The decision was almost forced upon me by the fact that some of the problems I shall discuss are very controversial. For example, I shall explain in a later lecture why some experts (who at the moment have gone into hibernation) declare that the average intelligence of Englishmen is almost certainly declining, and why other experts are almost sure that it is not. I shall take sides, because one opinion does seem to me to be weightier than the other; but what I shall *not* do is to discuss the consequences of any fall of intelligence.

Again, I shall put to you the purely theoretical problem of the limits of physical and intellectual improvement. Some biologists believe that, apart from certain recurrent accidents, a population can become uniform in all kinds of desirable 'inborn' qualities, and can maintain itself in that state of uniform excellence according to the simple formula that like begets like. Other biologists are inclined to think that inborn diversity or inequality is a necessary part of the texture of human populations, and that it is kept in being by means which are often incompatible with 'breeding true'. This problem lies at the very centre of eugenics, and I shall do my best to explain the difference of principle on which the argument turns.

In the most general terms, the questions I have in mind are these: can man go on evolving in the future as he has evolved in the past, or is there some good reason why his evolution should now have come to an end? What are the evolutionary forces acting on men today, and how far can we predict their effects? For example, it is often said that advances in medicine and hygiene are undermining the fitness of the human race. It is said, too, that the practice of having fewer children than one is capable of having is so unnatural that it is bound to have evil consequences, not excluding the ultimate extinction of mankind. Are these just gloomy philosophizings, or do they contain the elements of a most unwelcome truth? Is it even possible to predict and regulate the size of human populations, so that we do not start worrying about why the birth-rate is not going up almost as soon as we stop worrying about why it is not going down? This is the problem I shall consider in my present lecture; it is important, and it makes an excellent text for discussing the fallibility of prediction.

In the last of my six lectures I shall turn to still more general questions, and I shall discuss the sense and significance of the belief that man is beginning to evolve in an entirely novel way.

I think that answers to questions of this kind, in so far as it is possible to answer them, are deeply necessary for any understanding of the future of man; and when I say that they are necessary, please remember that I have not said, and do not imply, that they are sufficient. This is all I can say by way of excuse for leaving out so much that is promised by the title these lectures bear. Even so, I cannot hope to be lucky enough to escape the charge that my approach is materialistic. I can neither deny this charge nor admit it, because 'materialism' is a word that has lost its power to convey an exact meaning; I can, however, resent it, because it is a word that has not yet lost its power to cause offence. To my mind, 'analytical' or 'exploratory' would be a better description. But instead of worrying about which word to use, let me give you an illustration of the way in which a particular problem in human biology has been approached.

It has been fairly general experience that the ratio of the births of boys to the births of girls goes up towards the end of or shortly after major wars.[1]* The ratio of boys to girls in Great Britain went up after the First World War and again in about the middle of the Second. One way to explain this is to say that it represents Nature's attempts to make up for the loss of men. Superior people smile at this explanation, perhaps forgetting that it contains the elements of very good sense. Many natural processes are self-regulating—are so adjusted that they can compensate for the effects of disturbance; war is indeed a disturbance, and we should expect it to bring any such power of self-regulation into force. This is a satisfying explanation because it *classifies* the phenomenon: we feel we now know the kind of thing that has been going on. But it marks the end of a train of thought instead of the beginning of an exploration, and even if it were true—which seems unlikely, if only because the wartime change in the sex ratio occurred in some countries which were not at war—it would still leave us wondering about the means by which the process of self-regulation achieved its effect.

* The original notes are to be found at the end of this chapter.

Most attempts to explain the wartime change in the sex ratio treat it as a special case of a much more general phenomenon that has nothing to do with war: the fact that older mothers have relatively fewer boys than younger mothers.[2] How is this more general phenomenon to be explained? The most popular explanation runs as follows. It is a fact that, from birth onwards, boys and men are more fragile or vulnerable than girls or women, in the sense that, whatever their ages may be, their chances of living one more year are, on the average, slightly less. If we now make the rather dubious assumption that this greater fragility of boys is true from conception until birth as well as from birth onwards, then the answer seems clear: for some reason younger mothers provide a better environment in the womb than older mothers. This lessens the slight disadvantage of being a boy baby and so makes it understandable that a higher proportion of boys should survive till birth.

On closer enquiry, however, it turns out that a mother's age, as such, does not account for the general fact that older mothers have relatively fewer boys; in any case, it could not account for the sharp rise in the ratio of new-born boys to new-born girls in Great Britain between 1941 and 1942, because the ages of mothers did not change as they should have changed if this simple explanation were true.

However, this is not the full story. It is true, on the whole, that the children of younger mothers are also earlier children; are first or second children, say, instead of third or fourth; but it is not invariably true, because a woman over thirty might be having her first child while a woman under thirty was having her last. The age of a mother must therefore be distinguished from her *parity*, that is, her rank in terms of the number of children she has already had. Moreover, the children of younger mothers are usually the children of younger fathers, so the father's age must be considered too. As I have said, the most recent work suggests that the age of the mother, as such, has not much to do with the sex ratio of new-born children; the age of the father certainly has, though for unknown reasons; and so, perhaps, has the mother's parity.[3]

During the later years of the war there was a slight increase in the proportion of first and second children and probably a slight decrease in their fathers' ages as well; yet neither change seems to have occurred on a scale that could account for the alteration in the sex ratio. But although the explanations that have been suggested, and others like them, turn out to be inadequate, the point is that this is the *kind* of explanation we should seek.

Let me give one other example of the analytical method because it illustrates a different point. On the average, children born between May and October seem to get slightly higher scores in intelligence tests than children born from November to April. Is it the season of conception or birth that somehow affects the intelligence of children, in so far as these tests can measure it; or is it the intelligence of the parents that influences the season of conception of the child? The second must surely be the explanation: for example, when one compares the average scores of winter and summer children who are brothers or sisters, the difference between them almost completely disappears.[4] If the phenomenon were not to be analysed in this fashion, there is no limit to the fancies we might build upon the mere correlation between intelligence and season of birth; we might even be tempted to think that the Signs of the Zodiac had something to do with the matter; that there must, after all, be something in what the astrologers say.

I should now like to turn to the deeply important problem of trying to predict and to regulate the size of a human population. As I have implied, it is an instructive paradox that we are usually oppressed either by the fact that the birth-rate is unduly high or by the fact that it is unduly low; and the world today is such that we can worry about both at once. For reasons I shall now try to explain, there is not much likelihood that we shall ever cease to be worried by the one problem or the other, though we can hope for long periods of respite in which we need not worry very much. Populations are potentially capable of growing at compound interest, but cannot in fact grow for any great length of time at any net rate of compound interest which is persistently above or below zero.[5] If the rate remains persistently below zero the population will die out (for that is what a negative rate implies); and if the rate stays persistently above zero the population will grow without limit and must eventually starve. No one contests these simple truths; people hold different opinions about the problem of over-population, but the differences are about its immediate urgency and about the tactics that should be adopted at this present time. But as my chief concern is with method, with the process of foretelling, I propose to discuss the problems of prediction and analysis that concern Great Britain and, to a greater or lesser degree, the rest of the Western world.

Before the war a number of highly skilled demographers said that if the prevailing patterns of birth-rates and death-rates were to continue then the population of most advanced industrial countries would go down steeply, in a matter of tens of years.[6] They pointed out—what seems obvious now, though it was far from obvious then, even to some biologists—that no comfort was to be got from the fact that the populations of most of these countries were still increasing, because the increase was mainly due to the success of ingenious modern ways of postponing the death of people beyond child-bearing age. As a separate consideration, various forecasts were made of the size of our own population at various intervals up to the year 2000. It is already possible to see that these predictions were systematically mistaken: they were all too low. One of them, putting the population of England and Wales below 30,000,000 in the year 2000, falls short by 20,000,000 of what the Registrar General now thinks of as a likely figure.

Before I discuss the shortcomings of these pre-war forecasts I do want to make it clear that they were expressly carried out as statistical exercises on the basis of a number of perfectly understandable assumptions, and that it would be a disaster if experts stopped making predictions of this degree of importance merely for fear of being wrong. Moreover, they were a big improvement, in point of method, on some of the forecasts or diagnoses made even a few years before. As late as 1930 an eminent foreign biologist declared that nothing could provide a more sensitive measure of the biological health of a population than the ratio of the annual numbers of births and deaths. When he turned the searchlight of this conception upon Great Britain he found no cause whatsoever for alarm. What makes his judgement so infuriating is that it happened to be more nearly right, in its general tendency, than estimates based upon reasoning incomparably more exact.

I suppose there were three main sources of error in these earlier predictions. The first was lack of information. In spite of its obvious importance, we in Great Britain did not begin to record the ages at which mothers bear their children until 1938, about ninety years after the need for information of this type had been explicitly foreseen; and we still do not record the age of the father. But, more than that, we need to know about the size of families, and how many families there are of each particular size, and how families are successively built up in each year after marriage. If pre-war demographers had had the kind of information that has since been provided by the Family Census of 1946 and the general census of 1951 they would have approached their problems in a different way: indeed, it was because of their insistence that the Family Census of 1946 was carried out.

A second source of error was to place too much confidence in the power of an index like the so-called 'net reproduction rate' to measure a population's biological fitness, its power to replace itself from one generation to the next. Not so very long ago a socially conscious person who heard mention of the net reproduction rate at once assumed a grave expression, which showed that he understood its import, and may have been intended to show that he knew exactly what it meant. It is, in fact, a measure of fertility which makes allowance for mortality—which does not assume, as cruder measures do, that everyone is lucky enough to live up to and right through the period of reproduction. It is usually based on the female population only, and only on female births, and it is arrived at by an arithmetic exercise which there is no need to describe.[7] Conceived in just those terms—as a well-defined computation which takes into account both gain by birth and loss by death—it is a good way of summarizing in one figure some of the more important information about the mortality and fertility that happens at that time to be in force. It was perfectly well understood that the computation itself gave one no authority to assume that fertility and mortality would not alter; but, when it is used as a measure of replacement, the net reproduction rate can only be as valid as our reasons are for thinking that fertility and mortality will in fact remain constant.

In real life the net reproduction rate fluctuates far more from year to year than one would expect of any index that professes to be a fundamental measurement of reproductive health. Between 1930 and 1940 the net reproduction rate in America, as in many European countries, was below unity, that is, below the level of exact replacement, one for one. In 1952 it reached the fantastic figure of 1.56, corresponding to growth by compound interest at the rate of 56 per cent per generation; but not even an American population could change in a few years from one whose future was looked at pensively to one which looked as if it would get completely out of hand. The net reproduction rate is extremely sensitive to changes in, for example, the ages at which women bear their children—to changes which need not be of great importance when thought of in terms of the span of a human reproductive life. But there is a more fundamental objection to using the net reproduction rate, or any other index like it, to predict what will happen in years ahead; to make it clear I must explain what is meant by a 'stable' population.

A population has not merely a size; it has also a structure; and to describe a population at any moment one must know not only its total number but how that total is built up of people of every different age. A 'stable' population is so called because it has a constant or steady age-structure or age-distribution, one which will not change so long as the rates of fertility and mortality remain constant. Unlike any real population I am aware of, it can reproduce its structure from one generation to the next, and even regenerate or restore itself if some upheaval like a war or depression should temporarily change its shape. A stable population grows at a constant net rate of compound interest which may, of course, be zero, so that births and deaths cancel each other, and the population stays constant in size as well as shape.[8] Stability can be achieved only if the same age-specific rates of mortality and fertility have been in force for something like 100 years. No large population has ever achieved such a stability, and it is not at all likely that it ever will. This is why students of populations wear censorious frowns when people talk, as they so often do, of 'stabilizing' the population of the world or of one country or another at any particular figure they may have in mind; for, short of tyranny, it is not at all clear how any such stability could be achieved.

When it is used for predictive purposes, the net reproduction rate can be thought of as a preview of what the population's rate of increase would eventually come to be if fertility and mortality remained constant long enough for a state of stability to be achieved. But if the population is not stable to begin with (in fact it will not be) then its composition by age and sex will certainly alter—not *in spite of* the fact that fertility remains the same but *because* it remains the same. The assumption that fertility will remain constant therefore implies that the population will change in structure; and these changes, in turn, make it likely—though not logically certain—that fertility itself will change. At the very least they will make us look anew at our reasons for assuming tht it would remain constant. The net reproduction rate cannot be used as if we were taking the nation's temperature; as if we were assessing its state of reproductive health. It has yet to be shown that any one index of fertility can be used for such a purpose.

All this sounds very disheartening; but out of the uneasiness and dissatisfaction of demographers a rather different style of analysis has emerged. The matter of principle involved is this. The life of a nation goes on from day to day and from year to year, and the changes that happen, historical changes, are marked against a scale of calendar time. But the lives and livelihoods and reproduction of men and women are marked against years of age, and the natural unit of demographical prediction is not, therefore, a calendar year or a subdivision of a century but a life or a subdivision of a life. The rates of fertility that will prevail here in five years' time will be shaped by today's teenagers, wondering when or whether they will get married; by people in their early twenties having their first children, and by people in their thirties having their later children or their last. Every such group has different experiences behind it and different prospects before it, but the fertility index we compute in five years' time will remain indifferent to them all. Yet it cannot be assumed that those who are twenty in 1965 will have the same fertility as the forty-year-olds had when *they* were twenty; or that they will grow up to have the fertility the forty-year-olds happened to have had in 1965.

The analysis of populations in terms of the changes that occur from one calendar year or decade to another is sometimes called 'secular analysis', and obviously it must be reinforced by analysis of a different style: one which takes all the people born in one year or married in one year and follows them through their lives. Analysis of this kind is called 'cohort analysis'—it slices time lengthways instead of cutting across it year by year. The adoption of cohort analysis is the most important advance in practical demography of the past ten years.[9] No one pretends that its adoption is an intellectual triumph; as I said, it was mainly lack of information that prevented its coming into use before; but in any empirical sense it has been highly informative and revealing.

Cohort analysis makes it easier to resolve fertility into factors— sex ratio, marriage rates, the ages at which people marry; above all, it has shown how important it is to know the pattern in which married couples build their families. No other method could have shown so clearly that the tremendous increase in the birth-rate which began towards the end of the war was mainly due to a change in the pattern of making families: people began to have in 1942 and 1943 the children they would normally have had two or three years beforehand. The postponement of births need not imply that families are going to be smaller than they otherwise would have been, and need not therefore have very much bearing on the problem of replacement. It is during times such as these, with changes in the rates and ages of marriage and in the pattern of building families, that indices like the net reproduction rate are least informative.

The most striking single fact that has emerged from cohort analysis is the remarkably unwavering trend of the size of completed families. It has fallen smoothly from an average of just over six for couples married in the 1860s to an average of just over two for those who married in the 1930s. There is an increasing element of guesswork in estimating the number of children of later marriages because not all of them have yet been born; cohort analysis can never be completely up to date. But there is a stability about the pattern of making families which suggests that forecasts founded upon cohort analysis are going to be nearer the mark than any made before the war.

As for replacement, I do not know that any demographer, on

present evidence, now fears a serious decline in the population of Great Britain. The latest estimates suggest that we are just about breaking even; demographers are perhaps temperamentally disinclined to put them higher, if only to correct the illusion that all must now be well because the birth-rate went up so rapidly after the war. There are signs, though, that the most recently married couples are going to have larger families; certainly the marriage rate has been going up and the average age at marriage going down*—although this does not imply that people who marry nowadays in their early twenties are going to have families of the same size as those who married in their early twenties before the war. In so far as purely biological pressures can influence marriage rates and ages, I guess that the present upward turn may be genuine and not just temporary. In my next lecture I shall refer to the fact that the average age at which children become sexually mature is still going down. Pressures of this kind may not be strong but they are very insistent; combined with everything that goes with a system of social security they could well increase fertility or, at least, change families to a pattern in which married couples have all the children they intend to have by an earlier age than hitherto. I should not be in the least surprised if in the nineteen-seventies or nineteen-eighties we in Great Britain were to start exchanging uneasy glances about the dangers of over-population, and wondering where things were going to end.

I have been saying that human lives, generation by generation, have a much longer stride than the march of history by calendar years or decades, so that it can be very misleading to assess the reproductive health or future size of a population from the fertility that prevails in any one year or group of years. The advantage of cohort analysis is that it makes it easier to resolve fertility into factors which have a meaning in terms of the way in which people actually behave. Predictions founded upon cohort analysis are somewhat more exact in the sense that one can foresee a little more clearly what follows from one's assumptions; and if these predictions are wrong, as to some extent they surely will be, it will be easier in retrospect

* The average age at marriage has since slightly increased.

to see which assumptions were faulty and which factors changed in unforeseeable ways.[10] This is about all that can be expected of predictions of this degree of complexity, though many biologists and demographers did at one time hope for more—to reveal in the growth of human population the unfolding of grand historical principles with the exigency and thrust of physical laws.

Furthermore, it is a technical error to suppose that in real life one can stabilize a human population, in the sense of bringing it to a state in which it will no longer change as a result of its own internal properties. Short of tyranny, all that can be done in an administrative sense is to coax and warn and bribe a population, to try to prevent its becoming unduly small; and to change these policies, with no sense of inconsistency or grievance, if it thereupon shows signs of becoming unduly large. In other words, policies can be adopted which fall equally far short of tyranny and of *laissez-faire*; they can be energetic and reasonable and effective without claiming to hold good in perpetuity or to be governed by the workings of grand demographic laws. I have a feeling that the same may be true, and true for much the same kind of reason, of other still more complex human affairs.

2. THE MEANING OF FITNESS

At the beginning of my first lecture I mentioned some of the questions I was hoping to answer, and among them were these two: Is there any real reason to suppose that advances in medicine and hygiene are undermining the fitness of the human race? And is man potentially capable of further evolution, or must we suppose that his evolution has now come to an end? In the course of trying to answer these questions I shall be obliged to use the words 'fitness', 'inheritance', and 'evolution', but to use them in narrow or unfamiliar ways. Scientists do sometimes tend to *brandish* these special usages at the layman, as if they had a sort of inner rightness; but it would be more gracious, and would reveal a better sense of language, if they apologized for them or explained them away.

In everyday speech 'fitness' means suitability or adaptedness or being in good condition; 'evolution' means gradual change, with the connotation of unfolding; and as for 'inheritance', we may hope to inherit money, rights, or property; we might inherit, too, a mother's eyes or a grandfather's gift for fiddling. These are the meanings (there is no need to say the 'proper' meanings) of fitness, evolution, and inheritance—the meanings for which scientists chose them when they were struggling to put their conceptions into words. In the course of time those conceptions have become clearer—more choosy, if you like—but the words which embody them have remained the same. The change that has gone on is sometimes described by saying that scientists give the meanings of words a new precision and refinement: a fair statement, were it not for the implication that they extract the true or pure meaning from crude ore. The innocent belief that words have an essential or inward meaning can lead to an appalling confusion and waste of time. Let us take it that our business is to attach words to ideas and definitions, not to attach definitions to words.[1]

The idea scientists now have in mind when they speak of 'fitness' can be explained like this. All the people alive 100 years from now will be our descendants, but not all of us will be their ancestors. In retrospect, therefore, it will be possible to give us scores or marks according to the share we took in being the ancestors of those future people; and those who took a larger share will be described as fitter than those who took a lesser share. The word 'fitness', then, has come to mean *net reproductive advantage*, and students of heredity, geneticists, do not deliberately use it in any other sense. One hears bitter complaints about this newer use of 'fitness', because it neglects so much of what is deeply important in human life: for example, the influence of good or evil people who happen to have no children but who are so obviously fit or unfit members of society in everything except this narrow genetic sense. But the contempt we may feel for the word must on no account be transferred to the idea that it embodies, an idea which has a central place in modern evolutionary thought.[2]

The argument that advances in medicine and hygiene are undermining the overall fitness of mankind is based on the belief that there is a hereditary or genetic element in all human ills and disabilities, even if it amounts to no more than a predisposition. This is known to be true of some diseases and not known to be false of any, so there can be no disagreement here. In its simplest form, the argument then runs as follows: because of the discovery of insulin, antibiotics, and so on, we are preserving for life and reproduction people who even ten years ago might have died. We are therefore preserving the genetically ill-favoured, the hereditary weaklings, who can intermarry with and therefore undermine the constitution of normal people; and as a result of all this, mankind is going downhill.

If by 'going downhill' is meant 'declining in biological fitness', with the implication that mankind will probably die out, this argument is simply a museum of self-contradictions.[3] It is true that we preserve for life people who even ten years ago might have died; but then we do not live ten years ago: we live today. It is also true that if some disaster were to destroy the great pharmaceutical industries to which diabetics and the victims of Addison's disease literally owe their lives, then a great many of them might die; but what could be deduced from this, except the lunatic inference that people who might conceivably die tomorrow might just as well be dead today? So let me put the argument in a form in which it might be put by a humane and intelligent person. He might say something like this:

I live in a country with a National Health Service, and the effect of this is that, in a sense, I myself suffer from diabetes and rheumatoid arthritis, and so on—from mental deficiency too. Of course my sufferings are only economic, in the sense that it is my taxes that help to pay the bill; but, as a result of them, I can afford to have only two children, though I very much wanted three. Now I am a sound and healthy person, and though I'm all for helping other less lucky people, it is clear that what you call their 'biological fitness' is being bought at the expense of mine.

There are two arguments here, and they cannot be considered apart. The first is that people of a genetically sound constitution are being crowded out by the inferior. My spokesman was too humane to resent the idea that the inferior should survive and have children, but he saw some danger in the fact that the population of the future would contain fewer of his descendants because it would contain more of theirs. The second point he makes is that inborn resistance to a disease can be taken as evidence of a *general* soundness of body, of fitness in some rounded and comprehensive sense; so that even if the people he described as unlucky could all be cured of their particular disabilities, there would still be a deep-seated, though hidden, deterioration of mankind.

The arguments I have just outlined are serious and respectable, but they are not generally valid; they may sometimes represent the very opposite of the truth. Consider one of the forms of inborn resistance to a very severe form of malaria, subtertian malaria. It is now known that people can enjoy a definite inborn resistance to subtertian malaria if their red blood corpuscles contain something between 30 and 40 per cent of an unusual form of haemoglobin, haemoglobin S as opposed to haemoglobin A. One of the consequences of possessing haemoglobin S is that the red blood cells tend to collapse if deprived of oxygen; they become sickle-shaped intead of remaining rounded, and people whose blood behaves in this way are said to show the 'sickle cell trait'. Sickle cell trait can be found in parts of Africa, in some Mediterranean countries, and in parts of India—always in places where malaria has been or still is rife. It is not a disabling condition, so its victims should not be said to 'suffer' from it; and, in any event, it confers a high degree of resistance to the multiplication of the malaria organism in the blood.

This sounds like a splendid example of Nature's ingenuity in coping with a particularly murderous disease, malaria, without the help of these new-fangled drugs; but until one knows the rest of the story one cannot appreciate how devilishly ingenious it is.

The formation of haemoglobin S instead of A is due to an inborn difference of a particularly uncompromising kind, in the sense that if a person is genetically qualified to produce haemoglobin S, by possessing the appropriate 'gene' or genetic factor, then he surely will. People who show sickle cell trait do so because they have inherited the gene that changes haemoglobin A to S from one, and only one, of their parents. But when two such people marry and have children, one quarter of their children, on the average, will inherit that gene from both their parents; all their haemoglobin, instead of only part of it, will be of type S; and as a result of this they usually die early in life of a destructive disease of the blood known as 'sickle cell anaemia'. This highly successful form of inborn resistance to malaria therefore makes it certain that a number of children will die.

The situation as a whole can be set out in the form of a balance sheet or equation. In some parts of the world where malaria is rife, people with sickle cell trait are the fittest people. Alongside them are, on the one hand, normal people, whose haemoglobin is wholly of type A; and, on the other hand, the victims of sickle cell anaemia, whose haemoglobin is wholly of type S. The proportion in which these three classes occur adjusts itself automatically to a pattern in which the loss of life due to malaria and to sickle cell anaemia nicely counterbalances the gain that is due to the greater fitness of those with sickle cell trait. Nevertheless, in malarial regions, populations which possess this genetic structure are fitter than populations which do not.[4]

Essentially the same explanation will account for the widespread occurrence in certain parts of Italy of the disease known as Cooley's anaemia and, not impossibly, for the otherwise paradoxically high incidence of a certain fatal inborn disease of the pancreas[5] in Great Britain and elsewhere. In all such cases it may turn out that there is, or recently has been, some special advantage in having inherited from one parent, and one parent only, the genetical factor which produces such disastrous effects when it is inherited from both.

The moral of this story—though morality seems to have little to do with it—is that mankind will improve if we stamp out inborn resistance to malaria by stamping out malaria itself. Sickle cell anaemia is in fact disappearing from the Negro population of America at about the rate we should expect if malaria had ceased to be a scourge to it 200 or 300 years ago. So the only good thing about inborn resistance to malaria is—inborn resistance to malaria: it does *not* reveal any general soundness of constitution; and this is just the opposite to what my imaginary spokesman supposed. It is simply not true to say that advances in medicine and hygiene must cause a genetical deterioration of mankind. There is more to be feared from a slow decline of human intelligence, but that is a different matter: *if* it is happening, it is because the rather stupid are biologically fitter than those who are innately more intelligent, not because medicine is striving to raise the biological fitness of those who might otherwise be hopelessly unfit.

This question of a possible decline of intelligence is very important, and I shall devote my fifth lecture to it; but, having referred to mind, and defects of mind, it is essential to make this point. Some forms of idiocy and imbecility are congenital. 'Congenital' is a vague word, but I use it here to refer to an idiocy which follows from an inborn defect of the genetic make-up as it was laid down at the moment of conception. This defect represents an inborn difference from other people, but it is no more a property of the genetic make-up *as a whole* than the inborn difference between people of blood-groups A and B. One particular form of imbecility, now known as phenylketonuria, is the effect of a single, particular, and accurately identified inborn error of metabolism. In point of inheritance it is essentially similar to another disturbance of metabolism, alkaptonuria, the most serious effect of which is usually no worse than a darkening of the urine after it is exposed to air. To suppose, then, that congenital imbecility pointed to some general inborn inadequacy or degeneracy is nonsense—ignorant and cruel nonsense, too. Our ambition should be to *cure* phenylket-onuria, for it is an illusion to suppose that congenital afflictions are necessarily incurable; and if eventually we do cure phenylketonuria, we shall in no sense be conniving at a genetical degradation of mankind.

In a later lecture I shall mention one form of gross mental defect, mongolism,* which is by no means so simple in origin as phenylketonuria: it is the result of a damaging genetic accident involving a whole chromosome, and it is not at all easy to see how it might be cured. But it *is* an accident, a particular accident, one which happens more often to the children of older mothers; it is not to be thought of as an outward fulfilment of some inner degeneracy of a family stock.[6] When you come to think of it, all defects of the genetic constitution must have an accidental or unpremeditated or casually intrusive quality—'epiphenomenal' is the word; for it is impossible, indeed self-contradictory, that any animal should have evolved into the possession of some complex and nicely balanced genetic make-up which rendered it unfit. It is this fact that justifies our always hoping to find a cure.

'Inheritance' was the second of the three words of which I said that biologists use them in special or unfamiliar ways. Just what is inherited when geneticists speak of inheritance? It is becoming increasingly popular to say that a child inherits certain genetical *instructions* about how his growth and development are to proceed. This sounds like an ordinary metaphor—very apt, no doubt, but perhaps misleading; but I do not think 'metaphor' is quite the word. The idea of genetical instruction has come into use because, under the influence of telephone engineers and higher mathematicians, we now recognize a general, abstract similarity between all kinds of different ways of transmitting information. The passage of genetical instructions from parents to children is a particular concrete example of the more general idea of an act of communication, and just as valid an example of that idea as the information which we transmit in writing, by telephone, or by direct speech.

A gramophone record is a solid object which contains instructions about what particular sounds a reproducing apparatus is to utter. Genetical instructions are also conveyed by solid objects, in this case chromosomes; and the specificity of the instructions—their

* Now known as Down's syndrome.

property of being this instruction and not that—is a specificity of chemical structure: different molecular patterns convey different information, just as different sinuosities of the grooves of a record embody instructions about making different sounds. The discovery that nucleic acids are the substances that embody genetical information is to my mind the most important discovery in modern science, but I shall not argue the point because nothing turns on whether you agree with me or not.[7]

Genetical instructions are sometimes strict and uncompromising, in the sense that they can be carried out in one way only, if at all. The nature of one's blood-group or haemoglobin is strictly governed by one's genetic constitution, with little or no opportunity for compromise. But much more often the instructions allow a certain latitude in their execution—we differ from one another partly because we received different genetical instructions from our parents, but partly because, from conception onwards, our surroundings have acted upon us differently, and have therefore affected the way in which those instructions are carried out.

The theory of evolution is a theory which declares that genetical instructions change in character in the course of time. In my first lecture I said that a population had a certain age-structure, revealed by classifying its members into groups by age. In the same way a population has a genetic structure, a particular pattern of genetic make-up; just as its members are of many different ages, so also are they of many different genetic kinds. *Evolution* is a change in the genetic structure of a population—but a systematic change, a change with a definite direction or consistent trend.

We have seen such changes happening in our own lifetimes. In many hospitals bacteria have come to resist the action of penicillin, because bacteria genetically qualified to develop that resistance, at one time a tiny minority, have become the prevailing type. Likewise the genetic structure of some populations of moths has altered: dark forms have now become the prevailing forms among the sooty vegetation of an industrial countryside.[8] For all we know to the contrary, changes of this kind are the rudiments of greater evolutionary changes; and it was Darwin's theory, you remember, that they have come about because the different members of a species are endowed with different degrees of fitness: they leave more, or fewer, descendants, as the case may be, and if this happens the genetic structure of the population as a whole must clearly change.

This is all very well as far as it goes, but if we are ever to get a complete understanding of evolution we must obviously try to arrive at a complete theory of inborn variation: what forms does it take, what makes it possible, how does it happen, how is it maintained?

An analytical theory of inborn variation is one which will explain it in terms of the properties of chromosomes and nucleic acids, the substances which convey and embody genetical instructions. Let me give an example of what I mean. We can be sure that, identical twins apart, each human being alive today differs genetically from any other human being; moreover, he is probably different from any other human being who has ever lived or is likely to live in thousands of years to come. The potential variation of human beings is enormously greater than their actual variation; to put it another way, the ratio of possible men to actual men is overwhelmingly large. What mechanisms provide for the stirring about and shuffling and recombining of genetical information that makes this virtually endless diversity possible?

The most ancient and perhaps the most fundamental mechanism or stratagem that serves this function is that which is known to geneticists, in one of its forms anyway, as 'crossing-over'. Crossing-over is the swapping of parts between two chromosomes—a process which can occur when they have a certain general correspondence of structure; the effect of it is to combine, or recombine, genetical information in novel ways. One day biochemists and biophysicists will tell us what properties of the chromosomes make it possible for this swapping to occur. Then again: mutation, the birth of a newly variant gene, is an important process in evolution: we must ask what property or properties of the materials of heredity make mutation possible. This is the *kind* of question we must ask if we are ever to understand the pattern and progress of evolution. Let me ask another such question. If we look back upon the course of evolution we can see that within many of its lines, within many grand pedigrees of descent, there has occurred a process of becoming more complex, or, as zoologists say, more 'advanced'. Mammals are more 'advanced' than fishes; insects are more 'advanced' than worms. In the long view there has been an increase in the complexity of the genetical instructions which, so to speak, authorize an animal to be whatever it is.

If we merely confine ourselves to talking about degrees of fitness, the process seems gratuitous; what properties of the hereditary material make it understandable that it should have occurred? An explanation can only be groped after, but one kind of explanation might run something like this. The molecules of nucleic acid are of the sort that chemists describe as 'polymeric': they repeat the general pattern of their structure lengthwise, and can therefore build upon themselves to increase in length. They have also the crucially important property of lending themselves to duplication, because after various chemical manœuvres two similar molecules can be formed where there was only one before. There are many other more subtle properties of this kind; for example, the ability to break up and rejoin, to increase in length by letting in new stretches between the ends. Taken all together, these properties amount to what might be called a *repetitiousness* of nucleic acids and chromosomes, a readiness to become manifold or luxuriant or to elaborate upon genetic information—it is difficult to know which word to use. It is of the physical nature of nucleic acids that they can offer up for selection even more complex sets of genetical instructions, can propose ever more complex solutions of the problem of remaining alive and reproducing. Every now and again one of these more complex solutions will be accepted, and so there is always a certain pre-existing inducement or authority for evolution to have what, in retrospect, we call an 'upward trend.'

All this is extremely lame and halting; but, as I said in my first lecture, it is a more useful way of trying to explain the phenomenon than by talking about a 'vital force' of some kind which inspires organisms to advance in evolutionary history. All I am asking is: what material properties of chromosomes and nucleic acids qualify them for the functions which they do in fact discharge?

However that may be, human beings are the outcome of a process which can perfectly well be described as an advancement; and the second of the two questions I put at the beginning of this lecture was, in effect, where could we go from here?

First let me say that, even in the last fifty years, profound changes have occurred in human populations which are certainly not evolutionary changes. In some countries, for example, the average rate of growth and development has been and still is steadily going up. In the Scandinavian countries the average age of onset of first menstruation has declined between four and six months per decade for the last seventy years. In this country the height of adolescent boys has gone up by about three-quarters of an inch per decade, and their greatest height is reached by about eighteen or nineteen, instead of by twenty-five or twenty-six. These changes have been brought about by better nurture and nourishment, particularly in the first five years of life. The well-fed class may be nearing the end of this process, but the average will continue to change until the less well-fed can catch them up.[9]

Other changes have happened in human history that might conceivably be evolutionary changes. There was once a music-hall joke of uncertain import which turned on what Mr Gladstone may or may not have said in (I hope I am right in saying) 1858. One of the things he *did* say or imply in 1858 was that colour vision may have developed in mankind since the days of Homer; for he told us that Homer's world, as Homer described it, was almost colourless; and he might have added that colour blindness does not seem to have been referred to in writing before 1684. But I understand that the poverty of colour words in Greek and other ancient languages is to be construed as lack of sensibility, not lack of sensitivity; as lack of perceptiveness, not of ability to perceive; and though full colour vision *might* have evolved within recorded history, there is no good evidence that it has done so.[10]

But evolutionary changes, as I defined them, have occurred repeatedly in human history. The rise and fall of the genetic factor responsible for sickle cell anaemia is one, and in later lectures I shall mention others. These are comparatively trivial changes: could man evolve *radically*, or be made to evolve radically, in future? I have left this question to the end because it is utterly pointless and distracting. The answer, to be delivered with every inflection of impatience, is yes indeed. The necessary conditions are satisfied: a luxuriance of inborn diversity, a system of mating that maintains it, and an unspecialized structure as the zoologist uses that word, a structure which does not in itself commit human beings to any one way of life. From the point of view of genetical evolution, human beings have retained an amateur status.

But in fussing over the nature of some great metamorphosis which might conceivably happen, but which could only happen in real life if we were to be the victims of a sustained and consistent tyranny tens of centuries long, we may forget to ask a really important question: what changes *are* happening in the genetic structure of human populations as a result of forces acting upon us now? I stand by my original decision not to attempt to predict these changes or to discuss their consequences. My question is, what kind of knowledge and understanding must we acquire about mankind, and about genetics generally, if we are to identify and predict such changes; and this is essentially what my next three lectures will be about. In my last lecture I shall give a still more cogent reason for saying that the question I put—*can* man evolve as animals may yet evolve?—is pointless, because he has in fact adopted a new kind of biological evolution (I emphasize, a biological evolution) to which a great deal of what I have said in this lecture does not apply.

3. THE LIMITS OF IMPROVEMENT

Fifty years of research into human genetics has made it clear that human beings abide by the same laws of heredity as other animals do. There are thousands of human pedigrees that illustrate our conformity to the Mendelian laws.[1] I shall not bother you with what these laws are; they are pretty well understood, and we 'obey' them in whatever sense other organisms may be said to obey them; but if we are to understand the genetical behaviour of a human *population* or of any other population we shall need to know a great deal more than that. We shall need to know in what way and to what degree the members of the population differ from one another, and how that diversity is maintained; to what extent inbreeding is practised, if at all; from how large a number and how big an area a mate may be chosen or lighted upon; and whether 'opposites attract each other' or whether like mates with like. We shall need to know how many chromosomes there are, and what may be the importance of the phenomenon called 'linkage' in keeping the genes on one chromosome together, or of crossing-over in letting them get apart. In short, we must try to understand the *genetic system* of human beings—not just the syntax of heredity but the whole of that which governs the flow of genetic information from one individual to another and from one generation to the next.[2] In this lecture I shall discuss the idea of a genetical system in rather general terms.

The genetic system of a species sets a limit to what it can do by way of evolving, but it is not a permanent fixture: it can itself evolve. The most important evolutionary change we ourselves have witnessed is the evolution, in hospitals, of strains of staphylococci and other bacteria which resist the action of penicillin and other antibiotic drugs. Suppose now that *antibiotin*—an antibacterial drug not yet discovered—were to come into use next year: I shall relax my self-imposed ban on soothsaying so far as to predict that, if it does so, we shall surely witness the evolution of resistant strains of bacteria.

I am saying, then, that bacteria have a genetic system which enables at least some of them to overcome misfortunes which have not yet happened, which even we ourselves cannot foresee. The idea of an organism's providing now for what may happen to it in an unknown future sounds paradoxical; but in fact makes perfectly straightforward sense. The bacteria with us today are the descendants of bacteria which, in the past, must have come through a whole succession of just such appalling hazards. Their ancestors must have come from populations which were variable enough to have contained the few odd members that could cope. The few bacteria that did cope were the ones that left descendants; but they carried with them a genetic system—a system of genetical habits, if you like—which made sure that their descendants would be as various and versatile as ever before. To say that bacteria evolved into a state of resistance on some one occasion is to tell only half the story: they must have come to possess the kind of genetic system that made it possible for that particular act of evolution to have occurred.

So much for the influence of men and their other enemies on bacteria. It is just possible that we might learn the same lesson from the effects of bacteria and *our* other enemies on *us*. In the past thousand years, we in Great Britain and in western Europe generally have had to cope with murderous irruptions of plague, leprosy, and the sweating-sickness; of great pox, small pox, diphtheria, cholera, tuberculosis, and influenza. We too, then, like bacteria, must often have been propagated through somewhat unrepresentative members of our kind—unrepresentative in the sense that, on any one occasion, those who survived an epidemic may have contained a high proportion of a genetically privileged minority. If this is a true account of the matter—and as we are still alive to wonder whether it is or not—it follows that human beings must have a genetic system which makes sure that the appropriate minorities do indeed exist.[3]

Coping with sudden attacks by infectious organisms is only one form of a problem that besets all living things: to provide not merely for *adaptedness* to the environment but for *adaptability*; to provide not only for what is happening now but for changes which, if the past is anything to go by, are all too likely to happen in the future. Moreover, it is not only a matter of the future. Although we cheerfully speak about *the* environment of an organism or a population, we know very well there is no such thing. A population of individuals lives in a range of environments, narrow or wide as the case may be; and adaptability is just as much a matter of being adapted to environments which differ from place to place as to environments which change from time to time.

In principle, we can imagine two extreme solutions of this problem of adaptability. The first would be to arrive at some one genetic constitution which endowed each individual with great versatility and great powers of accommodation and resistance, so that each one went forth into the world capable of coping with almost anything that might come its way. This constitution would have to be pretty faithfully reproduced from generation to generation; if it were not so—if when the animals bred together they produced a great variety of different offspring—then the genetic formula for being so adaptable would be lost, and the solution would lose its point.

The second solution would be to confer adaptability upon a *population* of animals without too nice a regard to the welfare and fate of its individual members, and this means adopting a genetic system with the very opposite property: one that provides for and maintains a great many inborn differences between one individual and another. If such a system were to be adopted, then, with luck, whatever happened, there would always be some members of the population who could survive and perpetuate their kind.[4]

In the past twenty years we have come to realize that most free-living organisms—perhaps all of them—adopt neither the one solution nor the other. As a biological enterprise the first has turned out to be too difficult, and the second is appallingly wasteful. What animals have adopted is a rather shifty compromise between the two.

What form does this compromise take in human beings? The answer we give to this question will colour all our thoughts about the genetic future of mankind. I feel, for example, that people who study eugenics are sometimes inclined to assume that man has adopted, or could adopt, the former of the two solutions; they have in the backs of their minds the idea of some one excellently well-adapted, all-round kind of human being who could be perpetuated according to the formula that 'like begets like': in other words, by 'breeding true'. What I shall do now, therefore, is to discuss the kinds of inborn diversity that prevail among human beings, to see to what extent they point to the adoption of one or other of the two extreme solutions I proposed.

The inborn differences between human beings seem to be of three main kinds. First, there are the differences that divide us into a great majority and a tiny minority. Nearly all of us are lucky enough not to have haemophilia, for example, or a disease like Huntington's chorea; only about one person in a quarter of a million suffers from the bizarre abnormality that makes the urine darken when exposed to air. It is true that with the departures from normality that are said to be 'fully recessive' in expression—that will not make themselves apparent unless the offending gene has been inherited from both parents—the people who carry the gene without giving evidence of it will greatly outnumber those who are actually afflicted by the disease; but, even so, it remains true to say that the great majority of us are neither the carriers of any one such harmful gene, nor the victims of its action.[5]

If this were the only kind of variation among human beings and other animals, the picture we should form in our minds of their genetic make-up would correspond to the first of the two solutions I proposed for the problem of adaptability. The commonest and fittest animal would be one that inherited a normal gene—a 'good' gene, let us call it—from both its parents, and transmitted it to all its offspring; it would be *homozygous*, as the saying is, and homozygous for nearly all its genes. The same would be true of almost every other animal it could mate with, so that its offspring would have almost exactly the same genetic make-up as itself.

How then would inborn diversity arise, and how could there be any evolution? The answer would run as follows. Unusual genes— new variants of the existing genes—arise repeatedly by the process of mutation. If the genetic make-up of an individual is as nicely adapted as we are assuming it to be, then these new genes that intrude themselves will tend to have bad effects: they will lower the fitness of their possessors; and if mutation were not, as it is known to be, a constantly recurring process, they would eventually die out. But every now and again a mutant gene would arise which conferred some advantage on its possessors; in time it would be given every opportunity to reveal its talents, because sexual reproduction, abetted by crossing-over and segregation and other genetical devices, would make sure that it was introduced into every different kind of genetic constitution the population could provide. If, in the outcome, it *did* confer an advantage, then the new gene would slowly displace the old one and become the predominant type, the normal, regular thing. While the new gene was being received into the Establishment, the population as a whole would obviously have to go through a stage in which the members who did or did not possess it were fifty-fifty; but this state of affairs would be temporary, and a portent of better things to come. According to this theory, then, inborn variety is maintained by the nagging pressure of recurrent mutation, and natural selection will almost always act in such a way as to preserve conformity, by weeding out the possessors of unusual or aberrant genes.

With some refinements I shall not go into, this was the conception that most of us had in mind as recently as twenty years ago: it is the classical conception of the elementary text books, the idea of a uniform, a predominantly homozygous population of well-adapted individuals whose offspring are almost always exactly like themselves. The idea was applied in practice to the breeding of livestock animals. Artificial selection, it was thought, could go on smoothly until it had used up all inborn diversity in respect of the characters for which the selection was being practised; and the breeder would end up with a uniform population which met his preconceived requirements and which could be relied upon to perpetuate itself by breeding true. There were some tiresome minor snags, to be sure, and also some major difficulties. If the classical conception were wholly true, why should inbreeding, which leads to uniformity, also lead to a loss of fitness, not uncommonly to extinction? But these difficulties could be explained away, sometimes convincingly.[6] For example, if one asked why artificial selection should so often lead to a serious loss of fitness, the answer seemed reasonable enough. Artificial selection, being an arbitrary process, is almost certain to upset some hardly won and nicely adjusted natural balance between the genes. Now the great difference between artificial and natural selection is this. When judging the effectiveness of natural selection, that is selection for fitness, we are always being wise after the event; with artificial selection we are trying to be wise before the event; and what the event proves is that we are all too often ignorant.

The real weaknesses of the classical conception arise not from its being untrue but from its professing to be the whole truth. Let us try to see how far it falls short of being true of men.

There are a number of characteristics which do *not* divide human beings into a huge majority who possess them and a tiny minority who possess some alternative variant instead. The property of belonging to blood groups A or B or AB or O divides us into distinct classes of which not one is an extreme minority. The same is true of most other blood groups, and of the factors which (because there are so many of them) make it useless in the long run to patch up one human being with a skin graft taken from another, unless the two should happen to be identical twins. Variations of this kind, in which there is no question of huge majorities and tiny minorities, are described as 'polymorphic', and polymorphic variation represents the second of the three kinds of ways in which I said that human beings differed. Many examples of polymorphism are known already, and a great many more are simply waiting to be discovered. Some of them are pretty ancient. Anthropoid apes have blood groups closely related to our own; and an all-star cast of exceptionally eminent geneticists was able to discern that, just like ourselves, some chimpanzees can taste and others cannot taste the compound phenylthiourea, which is extremely bitter, so I am told. So we are not dealing here with the temporary polymorphism that simply marks the ascent in the population of some newly favoured gene; nor, from what we know of the rarity of mutation, is it possible that polymorphism should be kept up by the occurrence of new mutations between one generation and the next.

At one time it was thought that polymorphism owed its common-ness to its utter triviality; sometimes it was important, to be sure, but only under circumstances which could be comfortably explained away. One's blood group seemed to be a matter of complete indifference *unless* one happened to need a blood transfusion, a contingency which Nature might be excused for having overlooked. Skin grafting can indeed show that we are all innately different, but what of it? It would indeed be a splendid thing if we could repair a severe burn with a skin graft from a voluntary donor; but burns are pretty well unheard of in Nature, and nothing could be more unnatural than the grafting by which we attempt their repair.

It is now certain that polymorphism is *not* a matter of indifference in any sense. Our subdivision into Rhesus-positive and Rhesus-negative blood groups is a qualifying condition for the occurrence of a destructive disease of the blood in new-born children—a disease which, in Great Britain, affects about one child in 150 and which caused about 400 deaths in 1957. People who are not of blood group O seem to enjoy some special protection against duodenal ulceration, particularly—and this is another polymorphism—if the chemical substances distinctive of the blood groups appear in their saliva and gastric juice as well as in their blood. People who can taste phenylthiourea seem to be slightly more liable to get one form of thyroid disease and slightly less liable to get another. These facts make little sense at present, but they do show that polymorphism is under some kind of pressure from natural selection.[7]

Polymorphism seems to arise from two main causes. The first is when, for any reason whatsoever, it is an advantage for the population to be subdivided into two or more distinct types which depend upon and therefore sustain each other. The most extreme example of this kind is the distinction between the sexes, and the mechanism that provides for *this* polymorphism has long been built into the genetic structure of higher organisms.

The other main cause of polymorphism is when, for any reason whatsoever, a so-called 'heterozygote' is fitter than a homozygote. By a 'heterozygote' I mean an organism that has a hybrid make-up with respect to some particular gene—that has inherited two different variants of a gene from its two parents instead of the same variant from both. I mentioned an example of this in my last lecture when I said that people whose blood contained a mixture of two different forms of haemoglobin. A and S, were more resistant to subtertian malaria than those who had only one. In this particular case a mixture is formed because its possessors are heterozygous with respect to one of the genes that govern the form of haemoglobin: instead of inheriting a gene of the same kind from both their parents, they inherited two different forms of it, one from each. And because such people are hybrids, they do not breed true. You may remember my saying that a quarter of the children of heterozygous parents have only one kind of haemoglobin, haemoglobin S. To complete the story I should add that a second quarter have the other type of haemoglobin, haemoglobin A; and the remaining half are heterozygotes like their parents.

Now consider what would follow if this state of affairs—a greater fitness of the heterozygote—were to be the universal rule. Almost everything that follows from what I called the 'classical' conception would have to be withdrawn. Natural selection would no longer be a force that makes for constancy and uniformity; on the contrary, it would *oblige* populations to remain diverse, because the heterozygotes would be favoured, and heterozygotes do not breed true. Mutation, so far from being the great source of inborn diversity, would be reduced to a very minor role. We should have to abandon the idea that the fittest organism could be fixed as the overwhelmingly predominant type in the population, because, being of hybrid constitution, it would always throw off variants inferior to itself. We should be constantly frustrated in our attempts to select and establish a uniform breed of livestock animal, and artificial selection would almost always be forced to a standstill when there was still plenty of inborn diversity in the population—but a diversity which, unfortunately, could not be used.[8]

The state of affairs I have just described is no more the whole truth than that which is envisaged by the 'classical' conception, but it is a greater part of the truth than we suspected twenty years ago. The origin of the superior fitness of the heterozygous constitution, when it is superior, is one of the mysteries of modern genetics. We can sometimes discern reasons why, in any one particular case, the heterozygote should be superior; but I can think of only one *general* reason why it should so often have tended to become so, and it is this: a free-living species whose members have to cope with environments which change from time to time and differ from place to place will tend to acquire a genetic system which forcibly maintains a certain pattern of genetic inequality or inborn variety. This is one possible solution of the problem of providing for adaptability—a solution to which most free-living organisms are to some extent committed. This argument may be quite mistaken: there may be no *general* reason why heterozygotes should often be the fittest organisms, or, if there is a general reason, it may not be the one I have outlined. But whether there is one reason or a multitude of particular reasons, there seems no doubt that some large part of human fitness is vested in a mechanism which provides for a high degree of genetic inequality and inborn diversity, which makes sure that there are plenty of different kinds of human beings; and this fact sets a limit to any purely theoretical fancies we may care to indulge in about the perfectibility of men.

I have left to the end the third of the three ways in which human beings may differ from one another, because I shall discuss it at greater length in my next two lectures. Most characteristics do *not* divide us into sharply distinct classes of the sort I have been discussing so far. Our heights, or wits, or, over the normal range of variation, our blood-pressures form a smoothly graded series; tallness or shortness, brightness or dullness, are simply stretches of a continuous range. The inheritance of differences of this kind behaves as if it were due to the co-operation and interaction of a very large number of genes; and the same goes for some characteristics that do necessarily divide us into distinct classes, like the number of children one can have or the number of hairs on one's head. Many differences that have a direct bearing on fitness are inherited in this fashion—differences of fecundity itself, for example, or growth-rate, or length of life.

Unhappily, the study of this form of inheritance, 'metrical' inheritance, is exceptionally difficult, both in theory and practice; but I think that research has gone far enough to reveal here, too, the workings of a compromise between the two extreme solutions of the problem of providing for adaptability.

I said earlier, you may remember, that if the classical conception represented the whole truth, a programme of artificial selection could proceed quite smoothly until all inborn diversity had been used up—a limit which would not be reached until the genes affecting the characters under selection had been fixed in their true-breeding, that is, their homozygous, form. If, on the other hand, animals of hybrid constitution were always the fittest, then the attempt to fix some desired kind of animal would be an uphill struggle, constantly opposed and usually frustrated by the fact that the fittest animals did not breed true. What then are the results of experiments on the selection of metrical characters? In general, steady progress is made to begin with, and the results begin to take shape as if the classical conception were true; but after a number of generations of selection, it usually becomes clear that something is going seriously wrong: the stock begins to deteriorate in fitness and may even die out. A limit to improvement is reached when there is still plenty of inborn variation, but variation of a kind that is not accessible or not amenable to selection. There is known to be more than one reason why this limit should be arrived at, but one important reason does indeed seem to be superior fitness conferred by the heterozygous make-up. Attempts at selection are, in fact, torn between conflicting interests: the characters we are hoping to establish and fix in the population—height or weight, perhaps, or, in the fruit-flies that are so often used for these experiments, bristliness—may well find their most extreme expression in the true-breeding homozygous form; but that is not going to be much consolation if these homozygous forms are inferior in fitness, and are therefore at a constant disadvantage compared with the forms that do not breed true. Artificial selection and natural selection pull opposite ways.[9]

The experiments which reveal the compromise I have been discussing were done on animals, but there is no reason at all to suppose that their results do not apply to men. Human beings, too, are to some extent committed to a genetic system which attaches a certain weight, perhaps great weight, to there being many different kinds of men. This state of affairs is part of a very ancient genetic heritage: it came about, perhaps, because no species of free-living animal which survives to give evidence on the matter can ever have achieved adaptedness by the total sacrifice of adaptability; and the maintenance of a high degree of inborn variety is one way, wasteful but biologically easy, by which adaptability can be achieved.[10]

Fortunately, there is now a new solution of the problem of providing for adaptability, and it goes some way towards making up for these inborn inequalities and imperfections of men which the older solution necessarily entails. This newer solution is to *improve the environment*, whether by a comparatively simple method like eradicating malaria or tuberculosis, or by the grander enterprise of attempting to cure all human ills and deficiencies. There is sound biological sense in this solution: Nature, hitherto, has been somewhat inept, and has reconciled herself to compromises; she can do better now. The extremely difficult and ingenious trick which has made it possible for human beings to adopt this newer solution I shall try to explain in my final lecture. For the present, I should like you to notice that it is the *humane* solution too ('humaneness', according to the dictionary, means 'characterized by such behaviour or disposition towards others as befits a man'). It will be important to contrast the picture I shall finally arrive at with that older social biology which said 'Three cheers for Natural Selection', 'the devil take the hindmost'—and much else about Nature's teeth and claws. As some biologists did at one time connive at the acceptance of this manifesto, I should perhaps mention that it is based upon a technical misunderstanding of Nature, of man's place in Nature, and of the nature of man.

4. THE GENETIC SYSTEM OF MAN

Sir Francis Galton, the founder of Eugenics, a humane and highly gifted man, tried to think of some reason why so many noble lineages should have died out for want of heirs. To be exact, he asked himself why twelve out of thirty-one peerages conferred upon English judges from the days of Queen Elizabeth had become extinct, and why six others had come perilously near extinction. The solution he hit upon was this. The peers had married heiresses. An heiress worth marrying for the size of her legacy must have come from a very small family: if she had had a brother or many sisters, her legacy could hardly have been so great. In the days when families were so much larger, a small family was usually a sign that the parents were unable to have more children—not a sign that they chose to have no more. Differences in the ability to have children are strongly inherited: somewhat infertile parents tend to have somewhat infertile children. And so the noble lines died out through an inheritance of infertility.[1]

Let us set aside improving thoughts about the evils of avarice and of a desire for self-aggrandisement, and draw only this particular moral: that where monogamy prevails, the fertility of a marriage is limited by the less fertile partner: one partner sterile sterilizes both. This is one example of how social habits or conventions may affect the genetic welfare of mankind. Monogamous marriage is part of the genetic system of most human beings. In my previous lecture I described the genetic system of a species as the whole of that which affects the flow of genetic information from one individual to another and from one generation to the next, and I discussed the idea of a genetic system in very general terms. The time has now come to point more directly at some of the factors which may influence the genetic make-up of human populations.

It is extremely difficult to think of any social habit or act of legislation that has *no* genetic consequences. Penal, fiscal, social, moral, medical, political, or educational laws, schemes, treatments, habits, or observances will all make *some* mark on our genetic structure.[2] Advances in the application of medicine have changed the entire pattern of the forces of mortality. Taxation which falls unequally upon people with different innate endowments will affect the number of children they have, and therefore the genetic make-up of future generations. Differences of educability are to some extent inherited. Most educated parents want their children to enjoy their own advantages—a costly ambition, sometimes; and to achieve it they sometimes have fewer children than they might have wished. Migration will intermingle people with different genetic constitutions. It may be a matter of movement from village to village in an agricultural community, or it can take place on a colossal scale: between 1820 and 1950 some 40,000,000 foreigners entered the United States. Economic pressures can promote migration and political action can stop it; both, then, may have genetic effects.[3]

But intermixture does not follow inevitably from migration. By marrying mainly among themselves, people who observe the Jewish religion retain a certain racial distinctness: their blood groups are characteristically those of eastern Mediterranean peoples; their fingerprints belong to a distinctive family of patterns; they are almost exempt from some hereditary diseases and almost the only victims of others. Jewish people owe their integrity to a special form of *assortative mating*, the mating of like with like.[4] The tendency of people of the same educational standing to marry each other has to be kept firmly in mind when trying to weigh up the heritability of intelligence; and the same applies to marriages between people of the same temperament or physique. Assortative mating between people of the same economic or social standing will tend to produce a genetical stratification of the community, but migration from class to class will tend to break that differentiation down.

Inbreeding could be classified as a form of assortative mating. Inbreeding makes for genetic uniformity, and its effect upon species like ours, genetically adapted to outbreeding, is invariably bad. For example, very rare hereditary disorders turn up much more frequently among the children of marriages between first cousins than among the population as a whole.[5] As a general rule the Roman Catholic Church does not allow marriages between first cousins—

another example of how religious observances may have genetic consequences.

There are other less obvious but perhaps more important ways in which our social structure can affect our genetic make-up. A man may not always choose an occupation that will turn his inborn abilities to their best advantage, but he will often choose an occupation in which his disadvantages show up least. In children's comic papers, the scholarly youth can be identified not only by his earnest expression but by his spectacles and his spots. The inference we are presumably expected to draw is that long poring over books has deranged his vision and brought on an impurity of the blood; but I detect also the faint implication that his scholarly habits were brought on by his poor eyesight or his sickly constitution: it was because the lad could not take to football that he took to books. From the neolithic revolution onwards—perhaps for the past 10,000 years—the tendency of men to occupy the niches in a complex society for which their constitution fitted them, or failed to unfit them, must have played a very important part in the enrichment of mankind.[6]

Obviously, then, a multitude of social forces can affect the genetic make-up of human populations. But the great problem is: what is the nature and magnitude of their effects? Sometimes, in a modest way, we can predict them. Assortative mating will keep Negroes and Whites genetically apart in America or South Africa—though we hardly need a geneticist to tell us that. The consequences of inbreeding and of marriages between first cousins can be foretold in general terms. Our knowledge of simple Mendelian heredity in human beings makes it possible to forecast the kind of children that will issue from certain marriages, and our powers of prediction would be greatly strengthened if it were always possible to identify the potential parents of victims of so-called 'recessive diseases'.[7] Again, it is a fact that radiations increase the rate of mutation, and that nearly all mutant genes which are obtrusive in their action are harmful. So far as we understand its effects, therefore, radiation is indeed a genetic insult to mankind. We know enough, I think, to be able to say that hybridization between people of different races need not be expected to lead to an improvement, because both races will probably have adopted the well-balanced genetic constitution that matches their own environments; on the other hand, it might prove favourable in the long run because hybridization enriches diversity and might therefore produce a more versatile genetic structure than before. But we cannot be sure.

There is indeed an immense amount that we cannot be sure about. Yet, in spite of that, you will have heard it said that the fall of nations can be traced to genetic causes—for example, to the persistent infertility of a ruling class; you have been alarmed by insistent declarations that we are declining rapidly in intelligence; you have been stirred by the pronouncement that man can now control his own evolution; you have admired the easy assurance with which some people put a genetic interpretation upon differences of temperament and character—of the qualities that make for leadership or bravery or co-operativeness, or for economic success; and you have been taken aback by the confidence with which some experts will assert what other experts will just as confidently deny. The idea that a man's genetic constitution is not merely important but all-important, and that genetic knowledge is so far advanced that we can make nice judgements about our past or future genetic health, together add up to the doctrine or state of mind which, to be in the fashion of these things, should be called *geneticism*. Geneticism is the application to human affairs (or indeed to livestock breeding or natural selection or evolution) of a genetic knowledge or understanding which is assumed to be very much greater than it really is. It may surprise you to know that there is still no comprehensive theory of the improvement of livestock animals by selection. Geneticists have made great progress with their attempts to substitute sound genetic principles for an empiricism of dos and don'ts; but until those principles have been established, it might be as well to forbear from grandiose prophetic or retrospective utterances about the genetic welfare of mankind.

The same goes for our understanding of evolution. Twenty years ago it all seemed easy: with mutation as a source of diversity, with selection to pick and choose, and with a mainly homozygous make-up to be aimed at, all we were left to wonder about was why on earth evolution should be so slow. But we know now that natural populations are obstinately diverse in their genetic make-up, and that the devices which make them so are bound to make them rather resistant to evolutionary change. Our former complacency can be traced, I suppose, to an understandable fault of temperament: scientists tend not to ask themselves questions until they can see the rudiments of an answer in their minds. Embarrassing questions tend to remain unasked or, if asked, to be answered rudely. That is why I thought it important, in a previous lecture, to put an innocent question about the causes of evolutionary advancement. And here is another: why does so much of evolution lend itself to a belief in the inheritance of acquired characters? As I shall explain in my final lecture, belief in Lamarckism—in the idea that the environment can somehow issue genetic instructions to living organisms—is founded upon a misconception far wider than merely concerns genetics; but the question I put—how comes it that a Lamarckian style of inheritance should be so astutely imitated?—must still be asked, though I shall not have time at present to explain how an answer has been taking shape.[8]

In the years since the war, the study of selection and evolution has undergone an important change of emphasis, one which is highly relevant to our attempt to interpret the action of genetic forces on mankind. I should like now to give you a general idea of what has been going on.

In my last lecture I classified the inborn differences between human beings. First, you may remember, there are characteristics that divide us into sharply distinct categories: we either have or have not the particular defect which makes itself apparent as haemophilia or as phenylketonuria; our blood group is *either* A *or* AB *or* B *or* O. If all such differences were to be discovered—a great many have been— then each human being could be labelled and identified by his possessing a certain particular *combination* of alternative genes. Mr X, we should say, is of blood group A (as opposed to B) and of M (as opposed to N); he is Rhesus-positive (as opposed to Rhesus-negative)—and so on for all his characteristics, not merely the properties of his blood. This system of labelling would confirm us in a straightforwardly atomistic conception of evolution—in the belief that the units of inheritance and evolution were always certain particular things called 'genes'.[9]

But then I mentioned another kind of difference—that which divides us not into sharply distinct classes but by gradations or degrees. So with height or shape or intelligence; with fertility and growth rate and length of life; and presumably with character and temperament as well. I said that these smoothly graded differences behaved as if they were under the combined control of a very large number of particular genes. Inheritance of this kind is accordingly described as 'polygenic'. Inheritance of the other kind can be called 'segregative' because it separates us into clear and well-defined groups.

Segregative differences were the first to be analysed successfully and they are the first we learn about when we study genetics. A number of early attempts to investigate what we now call polygenic inheritance came to grief because the principle of Mendelian heredity had not yet come to light: the principle that the inborn differences between organisms do *not* tend to decay and disappear as a result of a *blending* of characteristics. On the contrary, Mendelian heredity works in such a way that the proportion of individuals having one genetic make-up or another can be assumed to remain constant from generation to generation unless something special happens to prevent its being so. This was a discovery of Newtonian stature, and it can be put, if we like, in a Newtonian style: the genetic structure of a randomly interbreeding population remains constant from generation to generation except in so far as some impressed force (like natural selection) is brought to bear upon it. Just as for Newton it was not motion but *change* of motion that called for a special explanation, so it is a change of genetic structure that demands to be explained. So far as we know, Mendelian inheritance, in this general sense of a conservation of inborn diversity, is a principle that applies to the whole of genetic inheritance; but it must on no account be confused with what is sometimes called 'Mendelizing heredity' or 'simple Mendelian inheritance'—the three-to-one ratios you may have heard of, or the comparatively simple rules that are relied upon when a matter of disputed parentage is brought to law. This simple Mendelian inheritance is something of a special case.

The historical order in which genetic discoveries were made, and the order in which we were taught them, incline us to forget an important truth: that polygenic inheritance is the general rule and that segregative inheritance is always unusual and sometimes freakish. What is so unusual about segregative inheritance is this: that the genes to which we attribute it have pretty well the same effects no matter what the rest of a man's genetic constitution may be. With the very rarest exceptions, the gene that turns Mr X into someone of blood group A will also turn Mr Y into someone of blood group A, no matter how different Mr X and Mr Y may be in other respects. This state of affairs, as I say, is unusual; and because it has this unusual character, the inheritance of differences between our blood groups can be seen to obey the simple Mendelian rules. And what I meant by the word 'freakish' was this. If some complex chain of chemical processes is necessary for the normal working of the body, then any mutation which breaks one link in that chain will cause a pretty obvious and far-reaching abnormality. A mutant gene of this kind will almost always make itself apparent, no matter what the rest of the constitution may be; and because its effect is both obtrusive and uncompromising, its inheritance will be seen to obey simple Mendelian rules. This is why so much of the human heredity that is studiable reads like a doctor's case-book; but the forms of heredity that can be seen to obey fairly simple rules are not a representative sample of heredity as a whole.

The trouble is that the whole way in which we think about genetics has grown up round these somewhat unusual cases; and, as a result of this, we are under a constant psychological pressure to think of polygenic inheritance as if it were just a highly complicated form of simple segregative inheritance—as if it were elementary Mendelism scored for a full symphony orchestra. But polygenic inheritance must be studied in its own right. Without doubt it has many regularities, analogous to those we see in ordinary segregative inheritance; but most of them have yet to be discovered. The study of polygenic inheritance and of the effects of selection upon graded, measurable characters is still in its beginnings. Its analysis is time-consuming and very difficult. The people who study it are confronted by strange and at present inexplicable phenomena. They are the last people in the world to make cock-sure predictions about the consequences of selection—least of all about selection as it affects the welfare and destiny of man. It is not true that we now know how to control our own evolution—if by 'control' is meant directing it towards a predetermined goal. We are *not* entitled by our present knowledge to put a genetical construction upon the rise and fall of nations. We do *not* understand the inheritance of differences of temperament or character; all we do know about the matter is what we have learnt from the evolution of tameness or docility in domesticated animals: that some such characters are under some kind of genetical control.[10]

This does not mean that we have to preserve a self-righteous silence until further bulletins are issued from genetical back rooms. We can already use to very good purpose our knowledge of human medical genetics, which is mainly a genetics of segregative differences, for the reasons I have just explained. We can call attention to what seem to be dangers according to present ways of thinking—for example, to the decline of human intelligence which I shall discuss later. We can expose fallacies when they are cruel (like the belief that imbecility is the symptom of a general decay of the genetic constitution) or when they are merely silly, like the belief that some special virtue travels down the male (as opposed to the female) line of descent.[11] And then we can make cautious statements about the effects, so far as we understand them, of certain genetical practices or habits. One of the questions I put in my first lecture was this: is the practice of birth control and family limitation so unnatural that it is bound to have evil consequences? This is a question of great practical importance, so let me spend the rest of this lecture discussing what the consequences of birth control might be.

Two distinct things are involved in a deliberate limitation of the size of families, and they must be kept apart. The first is a restriction of the total number of children born to a married couple; the second is a tendency which need not go with it, though it usually does— a tendency to complete a family earlier in life than would otherwise have been the case.[12] I shall discuss the second first.

The most general effect of an earlier completion of families will be to shorten the average gap between successive generations: one generation will follow another more quickly, so that whatever genetical changes are happening will happen faster in terms of calendar years. More particularly, we can look forward to a sharp decline in the numbers of new-born children afflicted by mongolian idiocy or by any other disease that increases in frequency with their mother's age. We can expect the sex ratio at birth to shift still further in favour of males,* with the consequence that, for the first time perhaps, men will outnumber women in their marriageable years. We can expect, too, a decline in the frequency of ordinary (that is non-identical) twinning.[13]

A less obvious and much less important effect will be to relax the pressure of natural selection against inherited abnormalities which can defer their outward appearance until about the middle of life; I have in mind abnormalities like Huntington's chorea, manic depressive psychoses, and certain forms of cancer. On the whole, these diseases do not disqualify their victims from parenthood until *after* most of their children have been born—from which it follows that the genes responsible for them will have been passed on before natural selection has much opportunity to intercede. An earlier completion of families will reduce the force of selection still further—a matter of minor importance, perhaps, but not altogether negligible.[14]

So much seems fairly clear: the rest is a matter of guesswork. An earlier completion of families implies that a human pedigree will hereafter run through a succession of rather younger mothers— mothers much younger, on the average, than they would have been a hundred years ago. This may just conceivably be a good thing, for the following reason. It is now widely agreed that human egg-cells do not increase in number after birth: women make do with the entirely adequate number they have to begin with, and the egg-cells are used up progressively in course of life. This means that egg-cells are obliged to wait for years in the ovary before they are shed into the Fallopian tube, where they may or may not be fertilized. Too long

* There has in fact been a slight decline, with in 1988 104.8 males born for every 100 females.

a wait, repeated generation after generation, might just possibly be harmful. It would be very interesting to study the recorded pedigrees of noble families and trace the fate of the lineages that went through last daughters of last daughters of last daughters. I must tell you that when an experiment which reproduces this state of affairs is carried out on animals called *rotifers*, the lines of descent that pass repeatedly through older mothers invariably die out. But then rotifers are very lowly and highly aberrant pond animalcules; it would be most unsound to draw far-reaching conclusions from a comparison between rotifers and human beings of noble birth.[15]

A restriction of the total number of children has several genetic consequences, one of which is to make the work of human geneticists even more difficult than it already is. It will certainly reduce the frequency of all diseases of children—above all haemolytic disease of new-born children—which tend to take a more severe form or to occur more often in the later children of a family. But it has one effect which is genetically unfavourable, though on a rather microscopic scale. When married couples plan to have fewer children than they could have had, they can, and sometimes do, make up for the loss of one child by having another. If a child has been lost through the manifestation in it of some highly damaging recessive gene—of a gene that must be inherited from *both* parents if it is to make its presence felt—then there is a two-thirds chance that the normal child which might replace it will be a carrier, though not a victim, of that harmful gene. To some very slight degree, therefore, the effect of natural selection will be circumvented, for the harmful gene will circulate just a little more freely in the population than would otherwise have been the case.[16]

But what people really fear when they talk about the biological evils of birth control is this. In many countries, families are deliberately restricted to two or three or four. The pressure of natural selection against low degrees of fertility will therefore be, to some extent, relaxed. I mean that if families now average only two or three children, there will no longer be the same sharp discrimination between married couples who *could* only have had two or three children and those who, had they wished it, could have had ten or twelve. Some discrimination there surely will be; but it is theoretically possible that in a matter of tens or hundreds of generations the proportion of innately very fertile men and women may go down.

If this were to happen, I think it would be looked back upon as an example—yet another example—of the way in which the level of fertility comes to be adapted to the prevailing circumstances. It is a fallacy to assume, as I fear some biologists still do assume, that the fertility of a species is a kind of primeval fixture—as if animals and plants were driven by some demon of fertility to have vastly more offspring than are needed. One can hear it said that the explanation of natural selection itself is that living things produce an allegedly 'prodigious' number of offspring, of which only a chosen few are spared. But to say this is to forget that the level of fertility adopted by any species is just as much the *consequence* of natural selection as its cause. There is in fact no good reason to fear than an innate decline of human fertility must be a stage on the road to extinction or that we shall face a struggle to keep mankind alive.

Of course, one can imagine circumstances in which a low level of fertility might be very disadvantageous. Some frightful disaster might oblige a handful of human beings to populate the entire world anew. But why worry about the imaginary dangers of a low level of fertility in the distant future when confronted by the real dangers of a high level of fertility as it affects so many countries of the world today? These are real and present worries; yet one of them at least we can spare ourselves: on present evidence, there is no reason to believe that the world-wide adoption of the practice of birth control would have biologically malign effects. On the contrary, there is every reason to believe that failure to adopt some measure of family limitation will lead in the long run, to misery, privation, and economic distress.

5. INTELLIGENCE AND
FERTILITY

A cultured lady declares that now at last she understands why so many of El Greco's figures seem to us to be unnaturally tall and thin. It is because El Greco had a certain defect of vision which made him *see* people with that particular distortion; and as he saw them, so he drew them. But a child then pipes up with the following objection: 'Surely if his eyes made everything look too tall and thin, wouldn't he see his own pictures in a different way from us too, and wouldn't they look just as strange to him as they do to us? If *all* he was doing was painting people in the funny way he saw them, then surely his paintings would *have* to look all right to us if they were going to look all right to him.'

What an irritating child! But—what an intelligent one! There is a grasp here, an ability to reason, to follow an argument and detect its faults. These qualities, and others like them, add up to 'intelligence'; and in this lecture I propose to discuss the possibility that in some countries, Great Britain among them, the average level of human intelligence is going down.

If we classify children of any chosen age by the scores they get in intelligence tests, and then make a diagram showing what number or proportion were awarded each possible score, we shall find that the diagram is smooth and pretty well symmetrical. The average score divides the children equally, and the most numerous single group is the one whose members have this average and middling score. Moreover, the number of children who exceed the average by a certain quantity is about equal to the number who fall short of the average by that quantity; and as we get further and further from the average in either direction, so the number of children to be counted gets less and less.

What I have said about the distribution of scores in intelligence tests applies to other characteristics of human beings—to the heights of adult men or women, for example. Within the whole range of heights and wits there are people who are exceptional in the sense of being in a minority—uncommonly good at intelligence tests, or uncommonly small; but to call them 'abnormal' is a bit misleading, because it suggests that they are separated from the rest of us by sharp divisions or bold steps. It is true that some people do lie right outside the normal range of variation—are abnormally small or abnormally dull for reasons that call for special explanations. So it is with idiots and imbeciles: those who fall short of the average to the degree of utter incapability do seem to form a class apart.[1]

But it is no longer these unlucky people that biologists have in mind when they discuss the possibility that intelligence may be declining. Many years ago, to be sure, the rumour got around that mankind would lose its wits because idiots and imbeciles are riotously fertile.[2] In fact they are nothing of the kind. Many are sterile; and in any event confinement to home or homes makes it impossible for most of them to have children—a good example of what I discussed in my last lecture, the way in which legislation (in this case the laws of certification) can have genetic effects. No: the problem arises over the greater fertility of those who are somewhat below the average of intelligence; and the fear is that their progeny are tending to crowd the rest of the population out. This might happen for one or both of two reasons: because, generation by generation, they tend to have larger families than the more intelligent; or because, generation by generation, they tend to have them earlier in life. For if the more intelligent parents start having children later and space them more widely apart, then, even if they end up with the same size of family in the long run, they are bound to be left behind.

All rational discussion of the possibility that intelligence may be declining starts from our knowledge of a certain association between the average performance of children in intelligence tests and the size of the families they belong to: in some countries, Great Britain among them, children who belong to small families are known to do better in intelligence tests than the children of larger families. The relationship between the average score of children and the number of their brothers and sisters is pretty consistent over the whole range of family sizes: taken by and large, children with x brothers and sisters do better than children with $x + 1$. A great mass of evidence points to a clear *negative correlation* between the size of a family and the average performance of its members.[3]

Before asking how this negative correlation is to be explained, and what its genetic implications may be, we must take some view about what is to be inferred from a score in an intelligence test. Some people speak with angry contempt of '*so-called* intelligence tests'; having satisfied themselves of the absurdity of claims which psychologists no longer make for them (and which the better psychologists never did make), they dismiss the entire subject from their minds. Others profess to attach no meaning to the word 'intelligence'—but try calling them *un*intelligent and see how they react. At the risk of being peremptory, because time is short, I shall take the view that intelligence tests measure intellectual aptitudes which are important, though very far from all-important; and that these aptitudes make up a significant fraction of what we all of us call 'intelligence' in everyday life. Only one disclaimer is important: intelligence tests can be valuable when they are applied to children still at school and to feeble-minded adults; their application to adults in general is very much more restricted in scope.

There are quite a number of possible explanations of the negative correlation between intelligence and family size. One possibility is that, for some reason, a child's intelligence declines with his position in the family; the first child being the most intelligent; the last, the least. The intelligence of each child might depend upon the age of his mother when she bore him, for a mother must be older when she bears (say) her fourth child than when she bears her third. This idea goes against all common understanding, but for purely technical reasons it is rather difficult to test. If we set aside certain forms of imbecility which are obviously exceptional, the most accurate tests show that matters to do with rank of birth do *not* explain the relationship between intelligence and family size.[4]

A second possibility is that size of family can itself affect a child's proficiency in those intelligence tests which rely heavily upon some outward or inward skill in the use of words. For one thing, a child in a large family will listen and contribute much more to the unscholarly prattle of its brothers and sisters than will a child in a

family of two or three. There is good evidence that inexperience in
the use of words does play some part in the negative correlation
between scores in intelligence tests and family size. It seems entirely
reasonable that it should. Words are not merely the vehicles in which
thought is delivered: they are part of thinking; and lack of experience
in the use of words, even unspoken words, may well put a child at a
disadvantage in a test.[5]

There is a less direct way in which the size of the family he belongs
to might be related to a child's performance. On the whole children
of large families are not quite so tall, at any given age of childhood, as
the children of smaller families—perhaps because, on the average,
they have been a little less well nourished. They grow more slowly,
therefore, though they might make up for that disadvantage by
continuing their growth a little longer, ending up no smaller than the
better fed. But if, at any chronological age, the children of large
families are a little backward physically, might they not be backward
in mental growth as well? And may they not eventually catch up
with the others, given a little time? Mental and physical growth are
not exactly in gear, so backwardness in size can by no means be
construed as backwardness of mind; but so far as our meagre
evidence goes, there is some small but definite connection between
the intelligence of children and their size at any given age of childhood;
and, with some reservations, this might account for a certain small
part of the negative correlation between intelligence and family
size.[6]

Another possibility is that a lowly score in an intelligence test
is part of a child's inheritance from its parents, though not an
inheritance in the technical or genetic sense. Unintelligent parents,
we might reason, have large families because they have neither the
skill nor the will to have smaller families; and, being unintelligent,
their conversation and precepts will tend to have a rudely pragmatical
character, and their houses to be bare of books. The nature of the
home he comes from is known to affect a child's performance in
verbal tests of intelligence; but there is no suggestion here that a child
of unintelligent parents would be at any disadvantage if he were to be
brought up in a more educated home.

Yet another possibility is that the children of less intelligent parents do *not* start on the same footing as the children of the more intelligent; that their lack of intelligence is something which good upbringing can palliate but cannot completely cure; that differences of intelligence are inherited in the technical or genetic sense.

At this point I shall ask you to assume (what I think no one denies) that differences of intelligence *are* to some degree inborn. There are certain obstinate and persistent correlations of intelligence between parents and their children and between the children of a family among themselves—correlations that do not disappear when as much allowance as possible is made for differences of upbringing, environment, and family size. The study of these correlations in the population generally, and, in particular, of the values they take in foster-children and in identical twins who have been reared apart, suggests that no less than half of the observed variation of intelligence is an inborn variation. For many environments, it may be a good deal more than half. It does not do to be more particular, because the concept of an inborn variation in characters greatly affected by the environment is very complex, and the people who know most about it are the least inclined to express it in a numerically exact form.[7]

Let us agree, then, that differences of intelligence are strongly inherited. We must now ask, why do less intelligent parents run to larger families? Teachers, demographers, and social workers incline to believe that the answer is mainly this. Less intelligent parents have larger families because they are less well informed about birth control or less skilful in its practice; because they are less well able to see the material disadvantages of having more children than can be well provided for, or the more than material advantages of having the children one really wants. I do not like to put it this way because it seems to import a moral judgement which, valid or not, has no bearing on the argument. Someone might insist that it was *right* for all parents to have all the children they were capable of having, and that the unintelligent live up to that precept because, being more innocent than learned but worldly people, they have a clearer perception of what is right or wrong. All this is beside the point. The point is that they have more children, and are unintelligent, whether that does them credit or not; and if they do have more children, there is a certain presumption that innate intelligence in the population at large will decline. It should not decline at anything like the speed suggested by the boldly negative correlation between intelligence and size of family, because, as we have agreed, some part of that correlation can be traced to causes in which inborn differences of intelligence need play no part.

I say there is a *presumption* that the average level of intelligence will decline. It is not a certainty. In the first place, one highly important piece of information is missing. What about the intelligence of married couples who have no children, or of people who never marry at all? Those who are oppressed by the possibility of a decline of intelligence point, with some reason, at the many highly learned people who are childless; and they remind us that when a population is classified by the occupations of its members, something is to be learned from the fact that manual labourers are much more fertile than those who live mainly on their wits. But those who think that the dangers of a decline are greatly exaggerated point out that idiots and imbeciles, and some of the feeble-minded, are very infertile too. Our uncertainty about the intelligence of those who have no children is awkward because it means that we cannot give a very confident answer to a very important question: to what extent are the parents of each successive generation a representative sample of the population of which they form a part?

A second reason for saying that the decline of intelligence is no more than a presumption is that there can obviously be no certainty in the matter until we know exactly how differences of intelligence are inherited. The argument for a decline is based on the belief that differences of intelligence are under the control of a multitude of genes, no one of which can be recognized individually; and it is assumed that the contributions of these genes to intelligence are *additive* in a certain technical sense. Are these assumptions justifiable? There is no reason at all to doubt that inborn differences of intelligence over the normal range of variation are under the control of a very large number of genes; but the idea that their contributions are additive requires a little consideration.

The word 'additive' refers to a particular pattern of co-operation of interaction between genes. An additive pattern of interaction implies (amongst other things) that there will be no such thing as hybrid vigour in respect of intelligence; it implies that a person who is mainly heterozygous or hybrid in his make-up with respect to the many genes that control intelligence will lie somewhere between the extremes of brightness or dullness that correspond to those genes in their similar or homozygous forms.[8] If, on the contrary, the genes that controlled differences of intelligence were all to exert their greatest effect in the hybrid or heterozygous state, then there would be no correlation between the intelligences of parents and their children: the children of parents who both had the high intelligence conferred by a hybrid make-up would as often as not be of low intelligence, and parents of low intelligence would as often as not give birth to children much more intelligent than themselves. But in actual fact the correlation of intelligence between parents and children is just as great as the correlation between the children of a family among themselves, so there is no good reason to doubt that the genes interact in the manner which I have described as additive. Nor, as I say, is there any reason whatsoever to doubt that a great many genes are at work. Both assumptions I have made are therefore justifiable, and there is a fair case for the belief that intelligence is declining. There is an equally good case for the belief that the decline could not go on indefinitely, but this, for the moment, I shall defer.

Is there any *direct* evidence that intelligence is declining? In 1932, a grand survey was made of the scores in intelligence tests of 90,000 Scottish schoolchildren whose eleventh birthdays fell within that year. Fifteen years later, in 1947, a very similar test was carried out on some 70,000 children of the same ages. No decline was apparent: the boys' scores had improved slightly; the performance of girls had risen even more. Taken together, the children of 1947 were two or three months ahead of the children of 1932 in terms of mental age.[9]

At first sight these results were immensely reassuring. A decline of intelligence on the scale we now fear might not have been shown up by tests only fifteen years apart; but an increase was more than most people had dared to hope for. But could anything have happened to conceal a genuinely innate decline? Unhappily it could. The children of 1947 were, on the average, an inch and a half taller than their predecessors of 1932. They were ahead in physical as well as in mental age;[10] but this does not imply that they were bound to end up taller and brighter when they reached adult stature of body and mind. Again, the pattern of family sizes might have altered in those fifteen years. If it did, the alteration would surely have taken the form of a decline in the proportion of the largest families, and this alone would account for a certain rise of average score. There is much else besides. The children of 1932 were taught by methods that may not have prepared them so well for intelligence tests; they grew up when a wireless set had not yet become a voluble piece of ordinary household furniture, and they lacked, then, whatever experience in 'verbalization' (if you will pardon the expression) comes from a constant familiarity with the spoken word.

But still: I gave you theoretical reasons for thinking that there might be a slow decline of intelligence, and direct evidence which, taken at its face value, shows that no such decline occurred. Can the theory itself be incomplete or wrong?

Some geneticists believe that it is incomplete, and I should like to explain their reasons. I asked a moment ago why the less intelligent should run to larger families, and gave the answer that has seemed reasonable to most of us: they have larger families for reasons connected with their lack of intelligence. But some geneticists look to another explanation: that people of mediocre or rather lowly intelligence are intrinsically more fertile, innately more capable of having children, than people of very high or very low intelligence.

To accept this interpretation is by no means to deny that inborn differences of intelligence are controlled by a multitude of genes, or that inborn variation of intelligence is mainly additive in character. A new point is being made: that people of mainly heterozygous make-up are innately more fertile—are innately *fitter*, as biologists use that word. When people of mediocre intelligence marry and have children, then, in the simplest possible case, some half of their children will grow up to be like themselves; the other half will consist of relatively infertile people of very high and of very low intelligence in about equal numbers. The children of each successive generation will therefore be recruited mainly from parents of mediocre intelligence, but they will always include among them the very bright and the very dull. It is possible, as a theoretical exercise, to construct a balance sheet of intelligence in which gains and losses cancel each other out: a reservoir of parents of mediocre or even lowly intelligence maintains a natural and stable equilibrium in the population, for, among their children, the all but complete sterility of low-grade mental defectives will cancel with the lesser fertility of the very bright.[11]

There is one respect, I think, in which this argument carries a lot of weight. It sets a natural limit to any likely rise or fall of intelligence. If a tyrant were to carry out an experiment on human selection, in an attempt to raise the intelligence of all of us to its present maximum, or to degrade it to somewhere near the minimum that now prevails, then I feel sure that his attempts would be self-defeating: the population would dwindle in numbers and, in the extreme case, might die out. In the long run, the superior fitness of heterozygotes would frustrate his dastardly schemes.[12] This is a cheering hypothesis, but it does not imply, I fear, that our population is already in a state of equilibrium; that the average level of intelligence may not fall a good deal further yet. It does not imply that we have already used up all the resources of additive genetic variation that can be called upon before natural selection intervenes. Nothing could be more unrealistic than to suppose that our population is already in a state of natural and stable equilibrium, with a nice balance between gain of intelligence and loss. We cannot disregard the purely arbitrary element in whatever it is that decides the size of a family—disregard the massive evidence of the Royal Commission on Population on the spread of the practice of birth control.[13] Nor can we neglect the fact that habits of fertility keep changing rapidly. The census of 1911 revealed a sharp increase in the difference between the sizes of families born to labourers and to professional men, but there are hints in the census of 1951 that the difference may since have declined. There is no need to assume that professional men are innately more intelligent than labourers; the argument would be equally valid if for professional men and labourers we were to substitute the people who do or do not believe that intelligence will decline; I am saying that there has been a *change* in the habits of fertility, and that when such changes are in progress, the idea of a natural equilibrium must be set aside.

Much else could be said to the same effect. For example, it is not true that the most highly educated people are the least fertile. Apart from imbeciles and idiots, the least fertile members of our population in terms of educational standing are those whose schooling stopped short of university but went beyond what is legally required.[14] I feel that the members of this group are less fertile because they *choose* to have fewer children; I am not inclined to believe that they are either unusually intelligent or unusually stupid; so far as innate intelligence goes, they may be a perfectly fair sample of the population as a whole. Nor do I think that some subconscious premonition of infertility directs them towards occupations which they merely appear to choose. But they are a numerous class, and the least fertile; what they contribute to our understanding of a natural balance of fertility is evidence that no such balance exists.

Again, the pattern of mortality as it falls upon the large families of poorer people has changed dramatically in the past fifty years. The death-rate of children within a week or so of birth has fallen rapidly, and begins to compare with mortality in the children of the better off; but deaths during the first year of life—deaths due mainly to infectious diseases—have not yet fallen so far, and do not compare so well.[15] This is but a fragment of the evidence that must turn our thoughts away from the idea that we are in a state of equilibrium. At one time, I suppose, there may have been some natural equilibrium between intelligence and fertility—adjusted, perhaps, to an average family size of eight or ten. Perhaps matters were so adjusted that the brightest and dullest of our forebears were incapable of having families larger than four or five. But the families we have in mind today belong to the lower half of what is possible in the way of human fertility, and it is hard to believe that a new equilibrium could have grown up around families with an average size of two or three.

Let me now summarize this long and complicated argument. It is a fair assumption that a child's performance in an intelligence test— imperfect as such tests are—gives one *some* indication of its wits. It is a fact that the average performance of children in intelligence tests is related to the size of the family they belong to: the larger the number of their brothers and sisters, the lower, on the average, will be their scores. Part of this negative correlation between intelligence and size of family can be traced to causes which have no genetic implications, whether for good or ill. But differences of intelligence are strongly inherited, and in a manner which, in general terms, we think we understand. If innately unintelligent people tend to have larger families, then, with some qualifications, we can infer that the average level of intelligence will decline. There are good reasons for supposing that intelligence could not continue to decline indefinitely, but equally good reasons for thinking that it may have some way yet to go. In any event, the decline will be a slow one—much slower than the boldly negative correlation between intelligence and size of family might tempt one to suppose. All our conclusions on the matter fall very far short of certainty: there are serious weaknesses in our methods of analysis, and grave gaps in our knowledge which, it is to be hoped, someone will repair.[16]

Profound changes in habits of fertility have been taking place over the past fifty or hundred years; and they are not yet complete. The decline of intelligence (if indeed it is declining) may be a purely temporary phenomenon—a short-lived episode marking the slow transition from free reproduction accompanied by high mortality to restricted reproduction accompanied by low mortality. But even if the decline looked as if it might be long lasting, it would not be irremediable. Changes in the structure of taxation and in the award of family allowances and educational grants may already have removed some of the factors which have discouraged the more intelligent from having larger families; and in twenty-five years' time we may be laughing at our present misgivings. I do not, however, think that there is anything very much to be amused about just at present.

6. THE FUTURE OF MAN

In this last lecture, I shall discuss the origin in human beings of a new, a non-genetical, system of heredity and evolution based upon certain properties and activities of the brain. The existence of this non-genetical system of heredity is something you are perfectly well aware of. It was not biologists who first revealed to an incredulous world that human beings have brains; that having brains makes a lot of difference; and that a man may influence posterity by other than genetic means. Yet much of what I have read in the writings of biologists seems to say no more than this. I feel a biologist should contribute something towards our *understanding* of the distant origins of human tradition and behaviour, and this is what I shall now attempt. The attempt must be based upon hard thinking, as opposed to soft thinking; I mean, it must be thinking that covers ground and is based upon particulars, as opposed to that which finds its outlet in the mopings or exaltations of poetistic prose.

It will make my argument clearer if I build it upon an analogy. I should like you to consider an important difference between a juke-box and a gramophone—or, if you like, between a barrel-organ and a tape-recorder. A juke-box is an instrument which contains one or more gramophone records, one of which will play whatever is recorded upon it if a particular button is pressed. The act of pressing the button I shall describe as the 'stimulus'. The stimulus is specific: to each button there corresponds one record, and vice versa, so that there is a one-to-one relationship between stimulus and response. By pressing a button—any button—I am, in a sense, instructing the juke-box to play music; by pressing this button and not that, I am instructing it to play one piece of music and not another. But— I am not giving the juke-box *musical* instructions. The musical instructions are inscribed upon records that are part of the juke-box, not part of its environment: what a juke-box or barrel-organ can play on any one occasion depends upon structural or inbuilt properties of its own. I shall follow Professor Joshua Lederberg[1] in using the word 'elective' to describe the relationship between what the juke-box plays and the stimulus that impinges upon it from the outside world.

Now contrast this with a gramophone or any other reproducing apparatus. I have a gramophone, and one or more records somewhere in the environment outside it. To hear a particular piece of music, I go through certain motions with switches, and put a gramophone record on. As with the juke-box I am, in a sense, instructing the gramophone to play music, and a particular piece of music. But I am doing more than that: I am giving it musical instructions, inscribed in the grooves of the record I make it play. The gramophone itself contains no source of musical information, but the record reached the gramophone from the outside world. My relationship to the gramophone—again following Lederberg—I shall describe as 'instructive'; for, in a sense, I *taught* it what to play. With the juke-box, then—and the same goes for a musical-box or barrel-organ—the musical instructions are part of the system that responds to stimuli, and the stimuli are elective: they draw upon the inbuilt capabilities of the instrument. With a gramophone, and still more obviously with a tape recorder, the stimuli are instructive: they endow it with musical capabilities; they import into it musical information from the world outside.

It is we ourselves who have made juke-boxes and gramophones, and who decide what, if anything, they are to play. These facts are irrelevant to the analogy I have in mind, and can be forgotten from now on. Consider only the organism on the one hand—juke-box or gramophone; and, on the other hand, stimuli which impinge upon that organism from the world about it.

During the past ten years, biologists have come to realize that, by and large, organisms are very much more like juke-boxes than gramophones. Most of those reactions of organisms which we were formerly content to regard as instructive are in fact elective. The instructions an organism contains are not musical instructions inscribed in the grooves of a gramophone record, but *genetical* instructions embodied in chromosomes and nucleic acids. Let me give examples of what I mean.

The oldest example, and the most familiar, concerns the change that comes over a population of organisms when it undergoes an evolution. How should we classify the environmental stimuli that cause organisms to evolve? The Lamarckian theory, the theory that acquired characters can be inherited, is, in its most general form, an *instructive* theory of evolution. It declares that the environment can somehow issue genetical instructions to living organisms— instructions which, duly assimilated, can be passed on from one generation to the next. The blacksmith who is usually called upon to testify on these occasions gets mightily strong arms from forging; somehow this affects the cells that manufacture his spermatozoa, so that his children start life specially well able to develop strong arms. I have no time to explain our tremendous psychological inducement to believe in an instructive or Lamarckian theory of evolution, though in a somewhat more sophisticated form than this. I shall only say that every analysis of what has appeared to be a Lamarckian style of heredity has shown it to be *non*-Lamarckian.[2] So far as we know, the relationship between organism and environment in the evolutionary process is an elective relationship. The environment does *not* imprint genetical instructions upon living things.

Another example: bacteriologists have known for years that if bacteria are forced to live upon some new unfamiliar kind of foodstuff or are exposed to the action of an anti-bacterial drug, they acquire the ability to make use of that new food, or to make the drug harmless to them by breaking it down. The treatment was at one time referred to as the *training* of bacteria—with the clear implication that the new food or drug *taught* the bacteria how to manufacture the new ferments upon which their new behaviour depends. But it turns out that the process of training belies its name: it is not instructive. A bacterium can synthesize only those ferments which it is genetically entitled to synthesize. The process of training merely brings out or exploits or develops an innate potentiality of the bacterial population, a potentiality underwritten or subsidized by the particular genetic make-up of one or another of its members.[3]

The same argument probably applies to what goes on when animals develop. At one time there was great argument between 'preformationists' and those who believed in epigenesis. The pre-formationists declared that all development was an unfolding of something already there; the older extremists, whom we now laugh at, believed that a sperm was simply a miniature man. The doctrine of epigenesis, in an equally extreme form, declared that all organisms begin in a homogeneous state, with no apparent or actual structure; and that the embryo is moulded into its adult form solely by stimuli impinging upon it from outside. The truth lies somewhere between these two extreme conceptions. The genetic instructions are pre-formed, in the sense that they are already there, but their fulfilment is epigenetic—an interpretation that comes close to an elective theory of embryonic development. The environment brings out potentialities present in the embryo in a way which (as with the buttons on a juke-box) is exact and discriminating and specific; but it does not *instruct* the developing embryo in the manufacture of its particular ferments or proteins or whatever else it is made of. Those instructions are already embodied in the embryo: the environment causes them to be carried out.[4]

Until a year or two ago we all felt sure that *one* kind of behaviour indulged in by higher organisms did indeed depend upon the environment as a teacher or instructor. The entry or injection of a foreign substance into the tissues of an animal brings about an immunological reaction. The organism manufactures a specific protein, an 'antibody', which reacts upon the foreign substance, often in such a way as to prevent its doing harm. The formation of antibodies has a great deal to do with resistance to infectious disease. The relationship between a foreign substance and the particular antibody it evokes is exquisitely discriminating and specific; one human being can manufacture hundreds—conceivably thousands—of distinguishable antibodies, even against substances which have only recently been invented, like some of the synthetic chemicals used in industry or in the home. Is the reaction instructive or elective?—*surely*, we all felt, instructive. The organism learns from the chemical pattern of the invading substance just how a particular antibody should be assembled in an appropriate and distinctive way. Self-evident though this interpretation seems, many students of the matter are beginning to doubt it. They hold that the process of forming antibodies is probably elective in character.[5] The information which directs the synthesis of particular antibodies is part of the inbuilt genetical information of the cells that make them; the intruding foreign substance exploits that information and brings it out. It is the juke-box over again. I believe this theory is somewhere near the right one, though I do not accept some of the special constructions that have been put upon it.

So in spite of all will to believe otherwise, and for all that it seems to go against common sense, the picture we are forming of the organism is a juke-box picture—a juke-box containing genetical instructions inscribed upon chromosomes and nucleic acids in much the same kind of way as musical instructions are inscribed upon gramophone records. But what a triumph it would be if an organism could accept information from the environment—if the environment could be made to act in an instructive, not merely an elective, way! A few hundred million years ago a knowing visitor from another universe might have said: 'It's a splendid idea, and I see the point of it perfectly: it would solve—or could solve—the problems of adaptation, and make it possible for organisms to evolve in a much more efficient way than by natural selection. But it's far too difficult: it simply can't be done.'

But you know that it has been done, and that there is just one organ which can accept instruction from the environment: the brain. We know very little about it, but that in itself is evidence of how immensely complicated it is. The evolution of a brain was a feat of fantastic difficulty—the most spectacular enterprise since the origin of life itself. Yet the brain began, I suppose, as a device for responding to elective stimuli. *Instinctive* behaviour is behaviour in which the environment acts electively. If male sex hormones are deliberately injected into a hen, the hen will start behaving in male-like ways. The potentiality for behaving in a male-like manner must therefore have been present in the female; and by pressing (or, as students of behaviour usually say, 'releasing') the right button the environment can bring it out. But the higher parts of the brain respond to instructive stimuli: we *learn*.

Now let me carry the argument forward. It was a splendid idea to evolve into the possession of an organ that can respond to instructive stimuli, but the idea does not go far enough. If that were the whole story, we human beings might indeed live more successfully than other animals; but when we died, a new generation would have to start again from scratch. Let us go back for a moment to genetical instructions. A child at conception receives certain genetical instructions from its parents about how its growth and development are to proceed. Among these instructions there must be some which provide for the issue of further instructions; I mean, a child grows up in such a way that it, too, can eventually have children, and convey genetical instructions to them in turn. We are dealing here with a very special system of communication: a *hereditary* system. There are many examples of systems of this kind. A chain letter is perhaps the simplest: we receive a letter from a correspondent who asks us to write to a third party, asking him in turn to write a letter of the same kind to a fourth, and so on—a hereditary system. The most complicated example is provided by the human brain itself; for it does indeed act as intermediary in a hereditary system of its own. We do more than learn: we teach and hand on; tradition accumulates; we record information and wisdom in books.

Just as a hereditary system is a special kind of system of communication—one in which the instructions provide for the issue of further instructions—so there is a specially important kind of hereditary system: one in which the instructions passed on from one individual to another change in some systematic way in the course of time. A hereditary system with this property may be said to be conducting or undergoing an *evolution*. Genetic systems of heredity often transact evolutionary changes; so also does the hereditary system that is mediated through the brain. I think it is most important to distinguish between four stages in the evolution of a brain. The nervous system began, perhaps, as an organ which responded only to elective stimuli from the environment; the animal that possessed it reacted instinctively or by rote, if at all. There then arose a brain which could begin to accept instructive stimuli from the outside world; the brain in this sense has dim and hesitant beginnings going far back in geological time. The third stage, entirely distinguishable, was the evolution of a non-genetical system of heredity, founded upon the fact that the most complicated brains can do more than merely receive instructions; in one way or another they make it possible for the instructions to be handed on. The existence of this system of heredity—of tradition, in its most general sense—is a defining characteristic of human beings, and it has been important for, perhaps, 500,000 years. In the fourth stage, not clearly distinguishable from the third, there came about a systematic change in the nature of the instructions passed on from generation to generation—an evolution, therefore, and one which has been going at a great pace in the past 200 years. I shall borrow two words used for a slightly different purpose by the great demographer Alfred Lotka[6] to distinguish between the two systems of heredity enjoyed by man: *endosomatic* or internal heredity for the ordinary or genetical heredity we have in common with other animals; and *exosomatic* or external heredity for the non-genetic heredity that is peculiarly our own—the heredity that is mediated through tradition, by which I mean the transfer of information through non-genetic channels from one generation to the next.

I am, of course, saying something utterly obvious: society changes; we pass on knowledge and skills and understanding from one person to another and from one generation to the next; a man can indeed influence posterity by other than genetic means. But I wanted to put the matter in a way which shows that we must not distinguish a strictly biological evolution from a social, cultural, or technological evolution: *both* are biological evolutions: the distinction between them is that the one is genetical and the other is not.

What, then, is to be inferred from all this? What lessons are to be learned from the similarities and correspondences between the two systems of biological heredity possessed by human beings? The answer is important, and I shall now try to justify it: the answer, I believe, is almost none.

It is true that a number of amusing (but in one respect highly dangerous) parallels can be drawn between our two forms of heredity and evolution. Just as biologists speak in a kind of shorthand about the 'evolution' of hearts or ears or legs—it is too clumsy and long-winded to say every time that these organs participate in evolution, or are outward expressions of the course of evolution—so we can speak of the evolution of bicycles or wireless sets or aircraft with the same qualification in mind: they do not really evolve, but they are appendages, exosomatic organs if you like, that evolve with us.

And there are many correspondences between the two kinds of evolution. Both are gradual if we take the long view; but on closer inspection we shall find that novelties arise, not everywhere simultaneously—pneumatic tyres did not suddenly appear in the whole population of bicycles—but in a few members of the population: and if these novelties confer economic fitness, or fitness in some more ordinary and obvious sense, then the objects that possess them will spread through the population as a whole and become the prevailing types. In both styles of evolution we can witness an adaptive radiation, a deployment into different environments: there are wireless sets not only for the home, but for use in motor-cars or for carrying about. Some great dynasties die out—airships, for example, in common with the dinosaurs they were so often likened to; others become fixed and stable: toothbrushes retained the same design and constitution for more than a hundred years. And, no matter what the cause of it, we can see in our exosomatic appendages something equivalent to vestigial organs; how else should we describe those functionless buttons on the cuffs of men's coats?

All this sounds harmless enough: why should I have called it dangerous? The danger is that by calling attention to the similarities, which are not profound, we may forget the *differences* between our two styles of heredity and evolution; and the differences between them are indeed profound. In their hunger for synthesis and systematization, the evolutionary philosophers of the nineteenth century[7] and some of their modern counterparts have missed the point: they thought that great lessons were to be learnt from similarities between Darwinian and social evolution; but it is from the differences that all the great lessons are to be learnt. For one thing, our newer style of evolution is Lamarckian in nature. The environment cannot imprint genetical information upon us, but it can and does imprint non-genetical information which we can and do pass on. Acquired characters are indeed inherited. The blacksmith was under an illusion if he supposed that his habits of life could impress themselves upon the genetic make-up of his children; but there is no doubting his ability to teach his children his trade, so that they can grow up to be as stalwart and skilful as himself. It is because this newer evolution is so obviously Lamarckian in character that we are under psychological pressure to believe that genetical evolution must be so too. But although one or two biologists are still feebly trying to graft a Lamarckian or instructive interpretation upon ordinary genetical evolution, they are not nearly so foolish or dangerous as those who have attempted to graft a Darwinian or purely elective interpretation upon the newer, non-genetical, evolution of mankind.

The conception I have just outlined is, I think, a liberating conception. It means that we can jettison all reasoning based upon the idea that changes in society happen in the style and under the pressures of ordinary genetic evolution; abandon any idea that the direction of social change is governed by laws other than laws which have at some time been the subject of human decisions or acts of mind. That competition between one man and another is a necessary part of the texture of society; that societies are organisms which grow and must inevitably die; that division of labour within a society is akin to what we can see in colonies of insects; that the laws of genetics have an overriding authority; that social evolution has a direction forcibly imposed upon it by agencies beyond man's control—all these are biological judgements; but, I do assure you, bad judgements based upon a bad biology. In these lectures you will have noticed that I advocate a 'humane' solution of the problems of eugenics, particularly of the problems of those who have been handicapped by one or another manifestation of the ineptitude of nature. I have not claimed, and do not now claim, that humaneness is an attitude of mind enforced or authorized by some deep inner law of exosomatic heredity: there are technical reasons for supposing that no such laws can exist. I am not warning you against quack biology in order to set myself up as a rival pedlar of patent medicines. What I do say is that our policies and intentions are not to be based upon the supposition that Nature knows best; that we are at the mercy of natural laws, and flout them at our peril.

It is a profound truth—realized in the nineteenth century by only a handful of astute biologists and by philosophers hardly at all (indeed, most of those who held any views on the matter held a contrary opinion)—a profound truth that Nature does *not* know best; that genetical evolution, if we choose to look at it liverishly instead of with fatuous good humour, is a story of waste, makeshift, compromise, and blunder.

I could give a dozen illustrations of this judgement, but shall content myself with one. You will remember my referring to the immunological defences of the body, the reactions that are set in train by the invasion of the tissues by foreign substances. Reactions of this kind are more than important: they are essential. We can be sure of this because some unfortunate children almost completely lack the biochemical aptitude for making antibodies, the defensive substances upon which so much of resistance to infectious disease depends. Until a few years ago these children died, because only antibiotics like penicillin can keep them alive; for that reason, and because the chemical methods of identifying it have only recently been discovered, the disease I am referring to was only recognized in 1952.[8] The existence of this disease confirms us in our belief that the immunological defences are vitally important; but this does not mean that they are wonders of adaptation, as they are so often supposed to be. Our immunological defences are also an important source of injury, even of mortal injury.

For example: vertebrate animals evolved into the possession of immunological defences long before the coming of mammals. Mammals are viviparous: the young are nourished for some time within the body of the mother: and this (in some ways) admirable device raised for the first time in evolution the possibility that a mother might react immunologically upon her unborn children—might treat them as foreign bodies or as foreign grafts. The haemolytic disease that occurs in about one new-born child in 150 is an error of judgement of just this kind: it is, in effect, an immunological repudiation by the mother of her unborn child. Thus the existence of immunological reactions has not been fully reconciled with viviparity; and this is a blunder—the kind of blunder which, in human affairs, calls forth a question in the House, or even a strongly worded letter to *The Times*.

But this is only a fraction of the tale of woe. Anaphylactic shock, allergy, and hypersensitivity are all aberrations or miscarriages of the immunological process. Some infectious diseases are dangerous to us not because the body fails to defend itself against them but—

paradoxically—because it does defend itself: in a sense, the remedy *is* the disease. And within the past few years a new class of diseases has been identified, diseases which have it in common that the body can sometimes react upon its own constituents as if they were foreign to itself. Some diseases of the thyroid gland and some inflammatory diseases of nervous tissue belong to this category; rheumatoid arthritis, lupus erythematosus, and scleroderma may conceivably do so too.[9]* I say nothing about the accidents that used to occur in blood transfusions, immunological accidents; nor about the barriers, immunological barriers, that prevent our grafting skin from one person to another, useful though it would so often be; for transfusion and grafting are artificial processes, and, as I said in an earlier lecture, natural evolution cannot be reproached for failing to foresee what human beings might get up to. All I am concerned to show is that natural devices and dispositions are highly fallible. The immunological defences are dedicated to the proposition that anything foreign must be harmful; and this formula is ground out in a totally undiscriminating fashion with results that are sometimes irritating, sometimes harmful, and sometimes mortally harmful. It is far better to have immunological defences than not to have them; but this does not mean that we are to marvel at them as evidences of a high and wise design.

We can, then, improve upon nature; but the possibility of our doing so depends, very obviously, upon our continuing to explore into nature and to enlarge our knowledge and understanding of what is going on. If I were to argue the scientists' case, the case that exploration is a wise and sensible thing to do, I should try to convince you of it by particular reasoning and particular examples, each one of which could be discussed and weighed up; some, perhaps, to be found faulty. I should not say: Man is driven onwards by an exploratory instinct, and can only fulfil himself and his destiny by the ceaseless quest for Truth. As a matter of fact, animals do have what might be loosely called an inquisitiveness, an exploratory instinct;[10] but even if it were highly developed and extremely powerful, it would still not be binding upon us. We should not be *driven* to explore.

* They do.

Contrariwise, if someone were to plead the virtues of an intellectually pastoral existence, not merely quiet but acquiescent, and with no more than a pensive regret for not understanding what could have been understood; then I believe I could listen to his arguments and, if they were good ones, might even be convinced. But if he were to say that this course of action or inaction was the life that was authorized by Nature; that this was the life Nature provided for and intended us to lead; then I should tell him that he had no proper conception of Nature. People who brandish naturalistic principles at us are usually up to mischief. Think only of what we have suffered from a belief in the existence and overriding authority of a fighting instinct; from the doctrines of racial superiority and the metaphysics of blood and soil; from the belief that warfare between men or classes of men or nations represents a fulfilment of historical laws. These are all excuses of one kind or another, and pretty thin excuses. The inference we can draw from an analytical study of the differences between ourselves and other animals is surely this: that the bells which toll for mankind are—most of them, anyway—like the bells on Alpine cattle; they are attached to our own necks, and it must be *our* fault if they do not make·a cheerful and harmonious sound.

NOTES

Notes to Lecture 1

1. The sex ratio in England and Wales (live births of boys per 1,000 live births of girls) rose from 1,038 in the quinquennium 1911–15 to 1,051—the highest figure that had ever been recorded—in the quinquennium 1916–20. Between 1941 and 1942 it rose from 1,053 to 1,063, reaching 1,065 in 1944. Since then the sex ratio has centred upon 1,060. The sex ratio normally favours males, perhaps from conception onwards; the effect of war referred to in the text is to favour males still more.

   Demographical data of all kinds for England and Wales are summarized in the Registrar-General's annual *Statistical Review*, nowadays published in three parts: I, Medical; II, Civil; III, Commentary. Demographical data for the world generally are to be found in the annual *Demographic Yearbook* of the United Nations.

One comment on matters of terminology should be made without delay. Demographers use the words 'fertility' and 'fecundity' in the senses usually attached by biologists to 'fecundity' and 'fertility'. I have adopted the demographic usage: 'fertility', unqualified, means actual reproductive performance; where 'fecundity' or reproductive potential is intended, I use 'innate fertility', or words to that effect.

2. The sex ratio of children born in England and Wales in 1956 fell from 1,074 for children of mothers under twenty to 1,026 for children of mothers between forty and forty-four. Various fairly obvious catches prevent our taking this raw statistical observation at its face value, e.g. the facts that (*a*) a relatively high proportion of women under twenty are pregnant at marriage, so that the under-twenties may be a specially fertile group; and (*b*) the age-groupings are not equally representative of different occupational classes or different habits of birth-control, etc.

3. See, for example, J. Lejeune and R. Turpin, *C.R. Acad. Sci.*, Paris, 244, (1957), 1833; E. Novitsky and A. W. Kimball, *Amer. J. Human Genet.* 10 (1958), 268. For a 'physiological' rather than a demographical interpretation of the change in the sex ratio, see T. McKeown, *Proc. 1st Int. Congr. Human Genet.* 2 (1957), 382.

4. The analysis referred to was carried out by J. A. Fraser Roberts, *Brit. Med. J.* 1 (1944), 320.

5. This deliberately vague statement is about as near as we can get to a 'law' describing the actual growth of populations. The 'law of increase by compound interest' or exponential law of population growth is a hypothetical statement about the consequences of combining a real rate of fertility with an imaginary rate of mortality, i.e. a mortality assumed to be independent of the population's size.

6. For a retrospective analysis of a representative number of these forecasts, see P. R. Cox, *Demography* (2nd edn., Cambridge, 1957).

7. A *gross* reproduction rate (computed for females) is roughly speaking an answer to this question: 'How many new-born girls can this new-born girl be credited with, on the average, if she lives right through the period of child-bearing and has children at the average rates appropriate to each fraction of that period—rates known from current information about fertility in each age-group of the population as a whole?' The *net* reproduction rate takes into account the new-born girl's likelihood of living through the period of child-bearing, and weights the estimate accordingly. From English Life Table No. 11, of 1951, only 96.3 per

cent of new-born girls reach the age of twenty and only 93.8 per cent reach the age of forty. If a new-born girl can be credited, on the average, with just one new-born girl in the next generation, then the net reproduction rate is said to be unity and the population to be just holding its own. If the computational exercise is carried out for males instead of females, a different figure is arrived at; and this in itself is a source of some of the difficulties mentioned in the text.

The reorientation of our thought about the usefulness of computing reproduction rates of various kinds (there are many kinds) can be largely credited to J. Hajnal of the London School of Economics; see in particular *Population Studies*, I (1947–8), 137; *Reports and Selected Papers of the Statistics Committee*, 303 (*Papers of the Royal Commission on Population*, Vol. ii, London, HMSO, 1950); *1958 Annual Conf. Milbank Foundation*, (Milbank Memorial Fund, New York), p. 11. These papers contain a much more fundamental and thoroughgoing analysis than any attempted in my lecture. See also *The Determinants and Consequences of Population Trends* (United Nations, 1953).

8. The concept of 'stability' and the proof that a population subject to constant age-specific mortality and fertility will eventually adopt a stable structure is the work of A. J. Lotka: see *The Elements of Physical Biology* (Baltimore, Williams & Wilkins, 1925); *Théorie analytique des Associations biologiques*, Part II (Paris, Hermann, 1939). Lotka's solution of the integral equation which yields the 'true rate of natural increase' of a population has passed into the literature of genetics, usually without acknowledgement of its source.

9. Cohort analysis was used for the first time on any large scale by D. V. Glass and E. Grebenik in their report on the Family Census of 1946, *The Trend and Pattern of Fertility in Great Britain* (*Papers of the Royal Commission on Population*, Vol. vi, London, HMSO, 1954). It was also adopted in the larger and more recent *Fertility Report* on the General Census of 1951 (London, HMSO, 1959), a report which should be referred to for information on the pattern of building families. For American demographical data, see W. H. Grabill, C. V. Kiser, and P. K. Whelpton, *The Fertility of American Women* (New York, Wiley, 1958).

10. It is the apocalyptic style of forecasting one should beware of. For certain special purposes forecasting is essential; to answer this question, for example: 'Should our universities be expanded permanently to a size large enough to accommodate the great numbers of children born in the post-war years, or can that great number of births be treated as a

temporary bulge which temporary expedients can cope with?' There's no knowing. 'One future development . . . we *can* forecast with a good deal of confidence,' said the Royal Commission on Population in its Report to Parliament in 1949, 'namely, a substantial decline in the annual numbers of births over the next fifteen years.' Actually—and quite unforeseeably—the annual number of births began to rise sharply after 1955, and forecasts of the future sizes of universities have been revised accordingly. They are sure to be revised again.

Notes to Lecture 2

1. Discussions about (for example) the 'real' meanings of the words *living* and *dead* are felt to mark a low level in biological conversation. These words have no inner meaning which careful study will eventually disclose. Laymen use the word 'dead' to mean 'formerly alive'; they speak only fancifully of stones as dead and never of crystals living; but they can tell a living horse from a dead one and, what is more, can remember an apt metaphor that turns on the distinction. See N. W. Pirie, 'The Meaninglessness of the Terms Life and Living' (*Perspectives in Biochemistry*, ii, Cambridge, 1937).

2. The genetical usage of 'fitness' is an extreme attenuation of the ordinary usage; it is, in effect, a system of *pricing* the endowments of organisms in the currency of offspring, i.e. in terms of net reproductive performance. It is a genetic valuation of goods, not a statement about their nature or quality.

3. For if medical treatment confers fitness upon the unfit, there can be no fear of extinction; if it fails to do so, the fear of extinction does not arise. The 'going downhill' argument seems to contemplate the predicament of modern man in primitive surroundings, without insulin, penicillin, central heating, and other allegedly debilitating devices; but it is not clear why such an exercise should be supposed to be informative.

4. This interpretation of sickle cell trait is the outcome of a brilliant combined operation between geneticists, chemists, and clinicians, among them J. V. Neel, E. A. Beet, L. Pauling, V. M. Ingram, and A. C. Allison. For the theory that sickle cell trait confers resistance to subtertian malaria, see A. C. Allison, *Ann. Human Genet.* 19 (1954), 39; for the difference between haemoglobins A and S, see V. M. Ingram, *Brit. Med. Bull.* 15 (1957), 27. Accounts of the genetics of sickling can be found in J. A. Fraser Roberts, *An Introduction to Medical Genetics* (2nd edn.,

London, Oxford University Press, 1959); H. Kalmus, *Variation and Heredity* (London, Routledge & Kegan Paul, 1957). For phenylketonuria and alkaptonuria, referred to later in the lecture, see H. Harris, *Human Biochemical Genetics* (Cambridge, 1959).

5. Cystic disease of the pancreas, the frequency of which, in Great Britain, has been put at one in 2,000—a frequency much higher than mutation at known rates could account for. There has therefore grown up the uneasy suspicion that the carriers of the harmful gene may have, or may have had, some special advantage over normal people (see L. S. Penrose, 'Mutation in Man', in *The Effect of Radiation on Human Heredity* (WHO, Geneva, 1957), p. 101.

   Cooley's anaemia is now more generally known as thalassaemia, a form of anaemia formerly thought to be confined to the shores of the Mediterranean (hence the name). The disease has two manifestations: a milder *thalassaemia minor* in heterozygotes, i.e. those who inherit the offending gene from one parent only, and *thalassaemia major* in those (homozygotes) who inherit it from both. The frequency of thalassaemia in, for example, the Ferrara region of Italy is far higher than recurrent mutation could well account for; as with sickle cell trait, it is supposed that the heterozygotes, victims of thalassaemia minor, have had some advantage over normal people; but no one yet knows where that advantage lay. See G. Montalenti, *Atti IX Int. Cong. Genet.* 1 (1954), 554; J. V. Neel, *Proc. Xth Int. Cong. Genet.* 1 (1959), 108.

6. The fact that mongolism can be traced back to a particular chromosomal abnormality was discovered by J. Lejeune, M. Gauthier, and R. Turpin, *C.R. Acad. Sci., Paris*, 248 (1959), 602. Human beings have forty-six chromosomes—not, as we formerly believed, forty-eight (J. H. Tjio and A. Levan, *Hereditas*, 42 (1956), 1; C. E. Ford and J. L. Hamerton, *Nature*, 178 (1956), 1020); in mongols there is an extra chromosome, apparently because one pair exists in triplicate instead of in duplicate. The newer methods of studying human chromosomes devised by C. E. Ford and his colleagues at Harwell have already led to important progress in human genetics—in particular, to the identification of the chromosomal disorders that underlie various abnormalities of sexual development.

7. For many years the nucleic acids were thought of as a kind of stuffing or padding; no one quite knew what to make of them. Our present conceptions originate with the discovery by F. Bawden and N. W. Pirie that nucleic acid is an integral part of the molecules of some plant viruses, and by O. T. Avery and his colleagues that solutions of nucleic

acid can bring about a genetical transmutation in certain bacteria. A method by which genetical information can be encoded in nucleic acid was proposed by J. D. Watson and F. H. C. Crick, and their interpretation of the structure of nucleic acid is now generally accepted. The coding depends upon the sequential pattern, down the length of the molecule, of four different pairs of organic bases; it may be likened to a Morse code with four symbols instead of two.

8. Our knowledge of the evolution of 'industrial melanism' in moths derives from the work of E. B. Ford and his colleagues at Oxford.

9. See J. M. Tanner, *Growth at Adolescence* (Oxford, Blackwell, 1955). That growth rate and the rate of attainment of adolescence have been increasing is agreed upon by all parties, but a number of experts are still inclined to question the evidence which points to a secular increase in the height finally reached when growth stops. Direct information is not fully adequate; the fact that older adults are, on the average, shorter than young (but fully grown) adults might indeed be due to the fact that the older people were born before their young contemporaries and so belonged to an earlier and perhaps smaller generation; but it might also be due (*a*) to an actual shrinkage of individuals during their lifetimes, or (*b*) to a mortality biased against taller people, who would thus form an unrepresentatively small proportion of an older population. The balance of evidence does seem to me to turn in favour of a genuine secular increase in the heights of adults.

10. W. E. Gladstone, *Studies on Homer and the Homeric Age*, 3 (Oxford, 1858), 457. For a general discussion of Gladstone's and other opinions on the matter, see J. André, *Études sur les termes de couleur dans la langue latine* (Paris, Klincksieck, 1949).

Notes to Lecture 3

1. J. A. Fraser Roberts, *An Introduction to Medical Genetics* (2nd edn., London, Oxford University Press, 1959), contains a wealth of information on Mendelian inheritance in human beings, and, in addition, an introduction to the reasoning that underlies the interpretation of metrical inheritance. In his introduction to the study of human *Variation and Heredity* (London, Routledge & Kegan Paul, 1957), H. Kalmus deals with the 'genetic system' of mankind in exactly the sense intended in this lecture and the next; although I think his views on eugenics are sometimes unduly astringent (from a tendency to identify it with its

worst manifestations), and although I have not been convinced by his reasoning on one or two particular points, his humanely sceptical attitude towards the excesses of what I later call *geneticism* is entirely persuasive.

2. The idea of a genetic system which may itself be the subject of evolutionary change was first impressed upon biologists by C. D. Darlington, particularly in *The Evolution of Genetic Systems* (1st edn., Cambridge, 1939).

3. The importance of infectious diseases in the genetical transformations of mankind during the past few thousand years has been generally overlooked; for a general discussion of *Disease and Evolution*, see J. B. S. Haldane, *Ricerca Scientifica*, 19 (1949), 3.

4. C. H. Waddington's *The Strategy of the Genes* (London, Allen & Unwin, 1957) contains the most thorough analysis yet attempted of the concepts of adaptation and adaptability.

The antithesis as I have put it—between the adaptation of an individual on the one hand, and of a population on the other—rides roughshod over a great many subtleties of the interaction between an organism and its environment; but as a flat approximation to the distinction I had in mind, I think it will stand. Many biologists who have grasped the idea that it is a *population* (not a pedigree) that evolves do not yet realize that the product of evolution may also be a population—a population with a genetic structure shaped by natural selection, and with the genetical system which makes it possible for that structure to be maintained. Indeed, with free-living outbred populations the idea of a final *product* of evolution is itself misleading: such populations never stop evolving; but there are some elements in the genetic system of outbreeding organisms that could be thought of as devices for preventing too rapid a change of genetic structure in response to forces which may be purely temporary in their action.

5. The word 'one' in this sentence is important. The odds are heavily against any one individual's carrying any one named harmful gene in the heterozygous state; but without doubt each one of us carries *some* such genes. Estimates of an individual's average total load of damaging or lethal recessive genes—a matter of the utmost importance when attempting to weigh up the malign effects of radiations on mankind—range in number from three to eight: see H. M. Slatis, *Amer. J. Human Genet.* 6 (1954), 412; J. A Böök, *Ann. Human Genet.* 21 (1956), 191; N. E. Morton, J. F. Crow, and H. J. Muller, *Proc. Nat. Acad. Sci., Wash.* 42 (1956), 855. Paradoxically, the most rapidly lethal of these harmful

genes are the least harmful or most merciful; for, in the homozygous state, they cause the death of an embryo very early in life.

6. The 'classical' explanation of the depression consequent upon inbreeding runs as follows. Inbreeding leads to the fixation of genes in their homozygous forms. As luck will have it, the genes fixed in the homozygous state will sometimes be harmful or unwholesome, so that the inbred stock, if it survives at all (often it will not) will be at a disadvantage compared with wild outbred population in which harmful genes are masked by the dominance of 'wild type' genes. But it is rather unlikely that the *same* harmful genes will be fixed in two different inbred lines; if we cross two such lines, therefore, each can go some way towards making good the deficiencies of the other, and the hybrid stock will be more vigorous than either of the two parental stocks from which it was derived. This explanation is plausible, and it must surely represent some part of the truth; but it is not the whole truth.

Among the 'minor snags' referred to in the text is the vexatious fact that some characters sought after by breeders represent the action of genes in the heterozygous state (e.g. the blueness of blue Andalusian poultry or the roan coats of shorthorn cattle). Such characters cannot be 'fixed'.

7. On polymorphism generally, see E. B. Ford (e.g. *Nature*, 180 (1957) 1315), who has done most to clarify our thoughts about what should be properly described as polymorphism. K. Mather (*Evolution*, 9 (1955), 52) has explained how polymorphism might arise when two different genetic types in an interbreeding population are adapted to their environments in different ways but yet depend upon each other (as, in the simplest case, males and females do); he predicted that polymorphism might come about as a result of 'disruptive' selection, and his predictions have been borne out by experiment (J. M. Thoday, *Heredity*, 13 (1959), 187). It was R. A. Fisher, I believe, who first pointed out that polymorphism will arise when a heterozygote is favoured above the corresponding homozygotes.

Many examples of polymorphism were gathered together and discussed by J. S. Huxley in his Bateson Lecture (*Heredity*, 9 (1955), 1). For various aspects of polymorphism in man, especially in relation to the blood groups, see the reviews by J. A. Fraser Roberts (*Brit. Med. Bull.* 15 (1959), 129), and P. M. Sheppard (ibid., 134). That there might be some connection between disease of the thyroid and the ability or inability to taste phenylthiourea was first suggested by H. Harris, H. Kalmus and W. R. Trotter, and has since been confirmed by

F. D. Kitchin, W. Howel-Evans, C. A. Clarke, R. B. McConnell, and P. M. Sheppard (*Brit. Med. J.* 1 (1959), 1069).

8. The superior fitness conferred by heterozygous constitutions is sometimes referred to as 'heterosis' or 'hybrid vigour', but agreed meanings have not yet taken firm shape. Heterosis refers (or is assumed to refer) to a superior fitness conferred by heterozygosis at *particular* genetic loci, though it is understood that many such loci may be involved when, for example, two different inbred or partially inbred stocks are crossed. On no account should the concept of hybrid vigour be extended to *human* racial crosses, or to crosses between the members of two different wild populations belonging to the same species; for here the cross is made between two outbred and heterozygous populations, each one already to some extent adapted to its environment (see Lecture 4). For a general treatment of the problems of heterosis (including a contribution by L. S. Penrose on evidence of heterosis in man) see the proceedings of the discussion held at the Royal Society under the chairmanship of K. Mather (*Proc. Roy. Soc. B* 144 (1956), 143); see also a number of important publications by I. M. Lerner, of which the most recent is *The Genetic Basis of Selection* (New York, Wiley, 1958).

The idea that heterosis plays any large part in human fitness has been briskly contested by H. J. Muller (e.g. *Amer. J. Psychiatr.* 113 (1956), 481; *Bull. Amer. Math. Soc.* 64 (1958), 137; heterosis, he is inclined to believe, is a temporary state of affairs to be seen only in 'adaptations that have not yet stood the test of geological time'. I feel he is certainly right to challenge the *mystique* which has tended to grow up around the idea that heterozygosis as such is intrinsically laudable. Where heterozygotes are the fitter organisms, it can only be because the genetic system of an organism has become adjusted to that situation. We need not suppose that the adjustment is irreversible; as Mather has pointed out, animals whose genetic system is based upon inbreeding have evidently come to terms with homozygosis, though presumably at the expense of adaptability.

9. The thoroughgoing analysis of the behaviour of metrical characters under artificial selection is the work of the past ten years; for a clear insight into the many difficult problems involved, see J. Maynard Smith, *The Theory of Evolution* (London, Penguin Books, 1958). The account given in my lecture was necessarily condensed and oversimplified. Limits to improvement may be set (*a*) by using up all the variation that is *accessible* to selection until a general shaking-up of the genetic constitution discloses patterns of genetic combination which

had, until then, been obscured by linkage: see the classical paper by K. Mather and B. J. Harrison (*Heredity*, 3 (1949), 1; and (*b*) by using up all variation that is *amenable* to selection: for some fraction is virtually imprisoned by a greater fitness of heterozygotes. The part played by these and other factors has been the subject of intent research in the Department of Genetics, University of Birmingham; the Institute of Animal Genetics, University of Edinburgh, particularly by D. S. Falconer, E. C. Reeve, A. Robertson, and F. W. Robertson; and the Department of Zoology, University College, London. I do not think it is possible to make any general statement about which of the two factors mentioned above is the more important: it will clearly depend upon the genetic system of the species under investigation. Where the number of chromosomes is large, as it is in human beings, obstinate linkages may be the less important obstacle to improvement. For an important theoretical discussion of heterosis, see A. Robertson, *J. Genet.* 54 (1956), 236.

10. Much of my argument can be summarized by J. M. Thoday's epigram: 'The fit are those who fit their existing environments and whose descendants will fit future environments' (*A Century of Darwin*, ed. S. A. Barnett, (London, Heinemann, 1959), p. 317). In the main, the compromise between adaptation and adaptability has been well concealed: much research in the past ten years has shown that the members of hybrid populations of outbreeding animals are *more* uniform to outward appearance and more stable in their responses to the environment than any inbred or predominantly homozygous line derived from them (see J. Maynard Smith, op. cit.). Inborn diversity has therefore been reconciled to outward uniformity—and to such good effect that inborn uniformity may actually lead to an outward diversity. Such a state of affairs would be entirely paradoxical if the 'classical conception' referred to in the lecture were wholly true.

Notes to Lecture 4

1. F. Galton, *Hereditary Genius* (London, 1869). The disappearance of the noble lineages studied by Galton was perhaps too rapid to be explained merely by the hazard of random extinction.

2. See C. D. Darlington, *Nature*, 182 (1958), 14. As to taxation, I have been told that the structure of taxation in the Netherlands is such as to make marriage an alternative to destitution; but in this country there is a definite fiscal inducement to live in sin.

3. Human beings in sparsely populated agricultural or pastoral communities tend to form little genetic pools ('isolates') between which apparently capricious differences of genetic make-up may arise ('drift'): see, for example, B. Glass, *Amer. J. Phys. Anthropol.* 14 (1956), 451.

G. Dahlberg (*Genetics*, 14 (1929), 421) was the first to try to estimate the numerical sizes of human genetic isolates and to answer the question 'from how large a number of women, on the average, does a man choose his wife?' His methods (enterprising but somewhat unsound: see N. E. Morton, *Ann. Eugen.* 20 (1955), 116, gave a figure of the order of hundreds; more recent computations relying upon the same principle put the order of magnitude at thousands; at all events it is not a matter of tens of thousands, as optimistic young men may be tempted to assume. Geneticists with learnings towards anthropology are more interested nowadays in the spatial sizes of isolates: see the amusing analysis by L. Cavalli-Sforza (*Proc. Xth Int. Cong. Genet.* 1 (1959), 389) of the parish registers of the diocese of Parma, which contains some 300,000 souls engaged mainly in agricultural pursuits. There turns out to be a surprisingly regular relationship between the likelihood of marriage and the distance apart of the dwelling places of the future bride and groom. The relationship fits neatly with the hypothesis (fortunately unknown to Isaac Newton, whose metaphysical tendencies gained ground in his later years) that the attractive power of a person living at a distance r from a community of population-mass N is directly proportional to N and inversely proportional to the square of r. Sociologists have published several such analyses: J. H. S. Bossard (*Amer. J. Sociol.* 38 (1932), 219) classified 5,000 successive marriage licentiates in Philadelphia and found that more than half the bridegrooms lived within twenty blocks of their future brides; less than one-fifth lived in different cities.

4. For the fingerprints of Jews, see L. Sachs and M. Bat-Miriam, *Amer. J. Human Genet.* 9 (1957), 117. Infantile (not juvenile) amaurotic idiocy and pentosuria are far commoner in Jews than Gentiles, but cases of phenylketonuria among Jews are very rare.

The tendency of deaf-mutes to marry one another is one of the less familiar examples of assortative mating; their need for special training brings them together, and, beyond that, they have a special understanding of and sympathy for each other's needs. Assortative mating is said to have been not uncommon among albinos or dwarfs in the days when their exhibition in circus side-shows threw them together more often than would have come about by chance.

5. For the theory underlying the higher expectation of 'recessive diseases' among the children of marriages between first cousins, see J. A. Fraser Roberts, *An Introduction to Medical Genetics*, or H. Kalmus, *Variation and Heredity*. The phenomenon was well known to Sir Archibald Garrod, who called attention to the greater frequency of alkaptonuria among the children of cousin marriages. Much genetical use has been made of the fact that, in the Roman Catholic Church, cousin marriages can take place only by a recorded dispensation.

C. H. Allström (*Acta Genet. Statis. Med.* 8 (1958), 295) undertook a particularly thorough analysis of cousin marriages in Sweden. Marriages between first cousins were unconditionally forbidden until 1750; from then until 1844 they were allowed only by Royal dispensation, and each such dispensation was recorded. The frequency of cousin marriages in Sweden has not declined very greatly since then, and in some parts of Europe (e.g. Southern Italy, including Sicily) it seems to have risen.

6. J. B. S. Haldane and H. J. Muller are fond of pointing out the special contribution short-sighted people may have made to the welfare of primitive communities. The principle I mention is of fundamental importance and deserved a more serious-minded illustration. Various aspects of the matter have been discussed by R. F. Ewer (*New Biology*, 13 (1952), 117; W. H. Thorpe (*J. An. Ecol.* 14 (1945), 67; E. Schrödinger (*Mind and Matter* (Cambridge, 1958)); and C. H. Waddington (*Nature* 183 (1959), 1634. 'An animal by its behaviour', Waddington points out, 'contributes in a most important way to determining the nature and intensity of the selective pressures which will be exerted on it.'

7. The identification of the carriers of harmful recessive genes has made progress in recent years: see D. Y.-Y. Hsia, *Genetics*, 9 (1957), 98. For example, the carriers of the gene which (when inherited from both parents) is responsible for phenylketonuria can now be identified with fair accuracy: they are less well able than normal people to break down phenylalanine. About one-quarter of the children of a marriage between two such heterozygotes will be afflicted by phenylketonuria. One day, perhaps, people will be 'typed' for some of the harmful recessive genes they carry as often and as readily as they are nowadays grouped by the properties of their blood, and two carriers of the gene for phenylketonuria might well be warned of the possible consequences of their having children. 'Marriage counselling' of this kind seems both sensible and humane, but two qualifications should be borne in mind. (*a*) Any reduction in the incidence of phenylketonuria that may be brought about by discouraging marriages between heterozygotes will

be eugenic in a symptomatic sense, for it will in fact reduce the frequency of phenylketonuria; but it will not, of course, decrease the frequency of the offending gene. On the contrary, the frequency will rise, because natural selection (presumably exercised only against the overt sufferers, homozygotes) will be proportionately relaxed. See the discussion by J. Maynard Smith, *The Theory of Evolution*, pp. 302–5. (*b*) A reduction in the number of victims of phenylketonuria may, for want of subjects to investigate, postpone the discovery of a cure. This second point may be insubstantial, but I mention both to emphasize the fact that eugenics is by no means plain sailing, and that symptomatic and radical eugenics may sometimes be at cross purposes with each other.

8. C. H. Waddington expounds his important concept of *genetic assimilation* in *The Strategy of the Genes* (London, Allen & Unwin, 1957), and explains the Darwinian basis of ostensibly Lamarckian patterns of inheritance.

9. See, for example, I. M. Lerner, *The Genetic Basis of Selection* (New York, Wiley, 1958), pp. 20–1. It must be clearly understood that the word *gene* stands for a genetic, not a structural concept: a gene is known by its performance and not by its substantive properties. There is no a priori reason why the structural entity revealed by mutation should coincide exactly with that which is revealed by crossing over or by the exercise of a particular physiological action; see the particularly cogent analysis by G. Pontecorvo, *Trends in Genetic Analysis* (Oxford University Press, 1959). I suppose that the genetical definition of the units of inheritance will ultimately be superseded by a structural or molecular definition.

The systematic analysis of polygenic inheritance in the quantal language of Mendelian genetics is particularly associated with the name of K. Mather (*Biometrical Genetics* (London, Methuen, 1949)).

10. Here see H. Spurway, *J. Genet.* 53 (1955), 325. I cannot help feeling that some element of the tameness or docility of domesticated animals (e.g. sheep and cattle) is the product of selection for frank mental deficiency; and I wonder how tame rats compare for intelligence with wild rats.

11. A certain genealogist, hungry for 'scientific' evidence to justify the belief that the male line of descent is specially meritorious, swallowed the naïve belief that the Y-chromosome (which travels down the male line) contains the genetic determinants of virility. His reasoning is akin to that which is alleged to have induced Russian soldiers to steal electric light switches from German homes in order that they, too, should have electric light when they returned home. The sex chromosomes act essentially as switches which direct development into alternative

pathways. The genetic importance of the Y-chromosome as such varies from species to species—trivial, apparently, in fruit flies, though the study of various sexual anomalies shows that it has some importance in its own right in man (see *Lancet*, 1 (1959), 715). In fish, judicious experiments on selection have made it possible to shift the burden of sex-determination to chromosomes other than the 'sex' chromosomes.

12. Cohort analysis (see Lecture 1) makes it possible to work out what fraction of a whole family is completed in each year after the parents' marriage or the mother's birth. In this country (if we disregard the interruptions caused by two wars) there has been a slow progressive decline in the mean age at marriage and a slow increase in the proportion of the family completed by the fifth or tenth year after marriage. Facts and figures are to be found in the demographic documents referred to in the footnotes to Lecture 1. In this country those who married between 1900 and 1909 had 47 per cent of all their children within five years of marriage, and 74 per cent within ten. The corresponding figures for the 'marriage cohort' of 1925 were 55 per cent and 81 per cent respectively. In 1935, 57 per cent of all (first) marriages were contracted below twenty-five; in 1950, 70 per cent. The mean age at marriage (all the figures I quote are, of course, averages) fell by just over a year, for both men and women, between 1946 and 1955. There has, then, been a tendency towards an earlier completion of families; family limitation might have taken the form of a postponement of childbearing, but in this country it has not done so.

13. A decline in the incidence of mongolian idiocy follows directly from L. S. Penrose's demonstration that its frequency rises sharply with maternal age.

An interesting and up-to-date analysis of the frequency of mono-zygotic and dizygotic twinning as a function of maternal age is to be found in Part III of the Registrar-General's annual *Statistical Review* for 1956 (London, HMSO). As to the point about sex ratio, the Registrar-General's figures for the number of men per 1,000 women in the population of England and Wales in 1956 are as follows: age group 0–4, 1,052; 20–4, 1,027; 30–4, 996; 40–4, 970.

14. The form of cancer I refer to is that which often develops from (familial) intestinal polyposis. The inheritance of Huntington's chorea (for recent genetical investigations, see E. T. Reed and J. H. Chandler, *Amer. J. Human Genet.* 10 (1958), 201) is governed by a gene with a strong expression in the heterozygous state; the mean age of onset for both men and women has been put at thirty-five. Many years ago, G. Levit (*C.R.*

Acad. Sci. URSS, 2 (1935), 502) pointed out that genes such as those responsible for Huntington's chorea may be classified as recessive in expression for the first part of life and dominant thereafter; and that the expected evolution towards recessiveness has, in this case, taken the form of a postponement of the overt action of the gene. Some of the wider implications of this phenomenon have been discussed by L. S. Penrose (*Amer. J. Mental Deficiency*, 46 (1942), 453) and by myself (*The Uniqueness of the Individual* (London, Methuen, 1957)). If there are inherited differences in the ages of onset of Huntington's chorea and other diseases of somewhat late expression, the action of natural selection must be to postpone their appearance. At one time it was believed that diseases of this kind made their appearance earlier and earlier in life in each successive generation—Nature's way, we were assured, of ridding herself of the genetic incubus, for eventually the victims would be afflicted too early to breed. The concept of 'anticipation' (as the phenomenon is called: echoes of it remain in the term *dementia praecox*) has been completely discredited: see A. Lewis's Galton Lecture, *Eugen. Rev.* 50 (1958), 91.

15. The source of nearly all our actuarial knowledge of rotifers is A. I. Lansing: see A. Comfort, *The Biology of Senescence* (London, Routledge & Kegan Paul, 1956).

16. The phenomenon I refer to here is far from obvious, and a word of technical explanation is called for. Consider a recessive gene a of frequency p which in homozygous form causes death at age one. After one generation of random mating, the frequencies of the genotypes aa, Aa, and AA will be in the ratio $p^2:2pq:q^2$ where $q = 1 - p$. The fraction p^2 will die, and the frequency of a in the residual population will therefore fall from p to $pq/(2pq + q^2)$. But if the members of the population are having fewer children than they could have had, they are in a position to replace the lost fraction p^2 of homozygotes by normal children of genotypes Aa or AA. The homozygotes, *ex hypothesi*, can arise only from marriages between heterozygotes ($Aa \times Aa$), from which it follows that two-thirds of the deficit p^2 will be made up by children of genotype Aa, and one-third by children of genotype AA. The frequencies of genotypes Aa and AA in the surviving population will now therefore be $2pq + \frac{2}{3}p^2$ and $q^2 + \frac{1}{3}p^2$ respectively; in other words, the frequency of a will now drop to only $pq + \frac{1}{3}p^2$. In fact, of course, a will be constantly reintroduced into the population by mutation, and a new equilibrium will be established in which a occurs with a frequency p' slightly higher than p. The effect at best (or at worst) is trivial; but it

might introduce yet another source of error into the 'indirect' method of computing mutation rates of recessive genes.

'Replacement', or compensation for lost children, is a real enough phenomenon: it may even amount to over-compensation (R. R. Race, *Ann. Eugen.* 11 (1942), 365). It applies not merely to a genetic situation of the kind described above, but also to the loss of children as a result of haemolytic disease (B. Glass, *Amer. J. Human Genet.* 2 (1950), 269). For a thorough formal analysis of the phenomenon, consult C. C. Li, *Amer. Nat.* 87 (1953), 257; R. C. Lewontin, ibid., 375.

Notes to Lecture 5

1. The distribution of scores in intelligence tests is, to a good approximation, normal (Gaussian): see J. A. Fraser Roberts on 'The Genetics of Oligophrenia', *Congr. Int. Psychiatrie* (Paris, Hermann, 1950), pp. 55–117. Fraser Roberts points out that whereas the feeble-minded form part of the lower end of the normal distribution of intelligence, idiots and imbeciles form a group that lies outside it.

2. For the history of popular attitudes towards fertility and mental abnormality, see A. Lewis, *Eugen. Rev.* 50 (1958), 91.

3. The case for a decline in the average level of intelligence, based upon the negative correlation between intelligence and size of family, is argued by C. Burt, *Intelligence and Fertility* (Occasional Papers on Eugenics, No. 2: London, Eugenics Society, 1952), and G. Thomson, *The Trend of National Intelligence* (ibid., No. 3, 1947). See also the memorandum by Thomson and the discussion arising out of it in Vol. v of the *Papers of the Royal Commission on Population* (London, HMSO, 1950). Most estimates of the magnitude of the correlation between intelligence and size of family put it between -0.2 and -0.3: see the admirable review by J. D. Nisbet, *Family Environment* (Occasional Papers on Eugenics, No. 8: London, Eugenics Society, 1953) and the discussion by A. Anastasi, *Psychol. Bull.* 53 (1956), 187). A particularly careful analytical study of intelligence and family size was carried out by J. A. Fraser Roberts, R. M. Norman, and R. Griffiths in the third of five important *Studies on a Child Population in Bath* (*Ann. Eugen.* 8 (1938), 178).

   The case for a decline of intelligence is no longer built upon demographic evidence that manual labourers, particularly unskilled labourers, are more prolific than scholars, administrators, and clerks: the facts are clear enough, but the inferences that used to be drawn from them are highly dubious. Nevertheless the illusion still persists that anyone who entertains the idea that intelligence may be declining is conniving at a fascist plot to discredit the working classes. Those who still hold this simple-minded view will be surprised by the temperateness of Thomson's and Burt's reasoning; but temperate though it may be, it is not free from genetical naïvetés—notably Burt's belief that the aptitudes revealed by intelligence tests are 'inborn' and virtually unin- fluenced by upbringing and environment, and Thomson's unawareness (understandable enough) of newer developments in the study of metrical inheritance.

4. See J. A. Fraser Roberts, *Brit. J. Psychol.* (Statistical Section (Oct. 1947), 35.

5. The argument here is J. D. Nisbet's, op. cit.

6. See J. M. Tanner, *Growth at Adolescence* (Oxford, Blackwell, 1955).

7. For a cogent discussion of the theory underlying the attempt to discrim- inate between genetic and environmental influences, see L. Hogben, *Nature and Nurture* (London, Allen & Unwin, 1945).

8. Not all geneticists use the word 'additive' in quite this sense: some confine it to interactions between alleles at different loci.

9. *The Trend of Scottish Intelligence*, Publ. Scottish Council for Research in Education, xxx (University of London Press, 1949). For the style of test used in 1947, see L. A. Terman and M. Merrill, *Measuring Intelligence* (London, Harrap, 1937).

10. The figure I mention—$1\frac{1}{2}$ inches—is arrived at by interpolation between two figures given to the Scottish Council by the Education Health Service of Glasgow: over the period 1932–47, the average height of nine-year-olds increased by 1.3 inches and of thirteen-year-olds by 1.7 inches: see *Social Implications of the 1947 Scottish Mental Survey*, Publ. Scottish Council for Research in Education, xxxv (University of London Press, 1953), the authors of which find it hard to believe that so great an increase of height (with all that it implies of better upbringing) should not be associated with an improved performance in intelligence tests. Dr J. M. Tanner has made the same point.

11. See L. S. Penrose, *Lancet*, 2 (1950), 425; *Brit. J. Psychol.* 40 (1950), 128; *Proc. Roy. Soc.* B 144 (1955), 203. Everyone is indebted to Penrose for importing some of the newer concepts of metrical inheritance and heterosis into the reasoning of educationalists and psychologists, but I disagree with his belief that the population of Great Britain can be supposed to be in a state of natural balance or genetic equilibrium. Human geneticists are fond of pointing out that human beings are a

genetically 'wild' population. One of the greatest handicaps to the genetical analysis of wild populations is the difficulty of obtaining reliable evidence about natural fertility and habits of mating. But there is a wealth of information of just this kind about human populations: if it exists, why not use it? That demographers and geneticists seem to live in worlds of their own is a puzzling anomaly.

12. See Lecture 3.

13. *Report of an Enquiry into Family Limitation* by E. Lewis-Faning. (*Papers of the Royal Commission on Population*, Vol. i, London, HMSO, 1949.)

14. See the *Fertility Report* (London, HMSO, 1959) on the General Census of 1951, particularly Table 3.4, p. xlix; the analysis is, of course, out of date, because it turns on a comparison between the family sizes of women aged forty-five to forty-nine at census and the family sizes of all women under fifty. There are hints from America that college-educated women now plan to have larger families than the less well educated: R. Freedman, P. K. Whelpton, and A. A. Campbell, *Family Planning, Sterility and Population Growth* (New York, McGraw Hill, 1959).

15. These remarks should be qualified by more recent evidence: see J. N. Morris, *Lancet*, 1 (1959), 303.

16. E. Hutchinson (*Amer. Nat.* 93 (1959), 81) has raised a problem which has several points in common with the one discussed here. It is a fair guess that there is some genetic element in abnormal sexual preferences or behaviour ('paraphilia', e.g. homosexuality), and that paraphilia is associated with some degree of infertility. How comes it, then, that paraphilia should not have been almost totally extinguished by the strong selection against it? Hutchinson inclines towards an explanation cognate with Penrose's (note 11 above): perhaps the most fertile people have a predominantly heterozygous make-up with respect to genetic factors affecting sexual behaviour; and perhaps paraphilia of one kind or another is the distinguishing mark of a mainly homozygous tail of the distribution of genotypes to be expected in the offspring of marriages between such heterozygotes. It will be many years before we can decide whether or not there is an element of truth in this interpretation. See also A. Comfort, ibid., 389.

Notes to Lecture 6

1. See J. Lederberg, *J. Cell. Comp. Physiol.* suppl. 1, 52 (1958), 398. It should not be necessary to say that a distinction such as that proposed by Lederberg can be apt and informative without claiming to hold good for

all time or to be unconditionally valid at all levels of analysis. Nor should it be necessary to point out that the analogy of juke-box versus gramophone is intended to do no more than guide the reader's thoughts towards the sense of the distinction.

2. Reviews by J. S. Huxley, *Evolution: The Modern Synthesis* (London, Allen & Unwin, 1942); P. B. Medawar, *The Uniqueness of the Individual* (London, Methuen, 1957).

3. The phenonemon described here should be distinguished from the evolution of resistant bacterial strains (see Lecture 2). The power of an *individual* bacterium to develop any particular 'adaptive enzyme' in response to a particular inductive stimulus depends upon its genetic constitution: the action of the 'inducer' is elective. Superimposed upon this phenomenon is the selection of those bacteria, among the population as a whole, which have the genetic constitution that enables them to respond in this way. It was this second phenomenon I referred to in Lecture 2.

4. The words 'evocator', 'releaser', and 'inductor', used by embryologists, students of behaviour, and bacteriologists respectively, all have the connotation of *elective* action in Lederberg's terminology. For some discussion of the embryological problem, see my 'Postscript' to *D'Arcy Wentworth Thompson*, by Ruth D'Arcy Thompson (Oxford University Press, 1958).

5. The 'elective' theory of antibody formation was the bold innovation of F. M. Burnet: see *The Clonal Selection Theory of Acquired Immunity* (Cambridge, 1959). One of the great problems confronting an elective theory is this: does the zygote from which an adult arises already contain enough genetical information to underwrite the synthesis of all the antibodies an adult is capable of forming? Or must we suppose that new genetical information comes into being during the course of development? Burnet inclines towards the latter interpretation, and in its present form his theory proposes that particular lineages of antibody-forming cells are qualified to produce only one kind of antibody. This, however, is only one construction that can be put upon an elective theory of antibody formation.

6. See A. J. Lotka, *Human Biol.* 17 (1945), 167. Lotka was thinking in particular of the 'evolution' of sensory and motor adjuncts: see my article on 'Tradition: the Evidence of Biology', in *The Uniqueness of the Individual*. I am extending it here to include cultural or social evolution in the wider sense envisaged by J. S. Huxley and C. H. Waddington.

7. See, for example, Karl Pearson's *The Grammar of Science* (1982).

Pearson was a humane man, and he struggles against what he believes to be the inescapable social implications of Darwinism; he could hardly bring himself to stomach the social Darwinism of, for example, Haeckel ('The theory of selection teaches us that in human life, exactly as in animal and plant life, at each place and time only a small privileged minority can continue to exist and flourish; the great mass must starve and more or less prematurely perish in misery. . . . We may deeply mourn this tragic fact, but we cannot deny or alter it!' The exclamation mark is mine.)

8. The disease is agammaglobulinaemia or, better, hypogammaglobu-linaemia. Most antibodies belong to the missing or almost missing gamma-globulin fraction of blood protein. See O. A. Bruton, *Pediatrics*, 9 (1952), 722.

9. The study of 'auto-immunity' and of autoimmune diseases has been particularly associated in recent years with the names of J. Freund and E. Witebsky. That Hashimoto's thyroiditis is essentially the consequence of a self-immunization was demonstrated by I. M. Roitt and D. Doniach. For a general review of these matters, see B. H. Waksman, *Experimental Allergic Encephalomyelitis and the Auto-Allergic Diseases* (Basle, Karger, 1959).

10. Discussed by S. A. Barnett, *Brit. J. Psychol.* 49 (1958), 289.

13

Osler's Razor
(1983)

Lewis Thomas is a physician, a scientist, a medical administrator, and a man of letters whose previous books, *The Lives of a Cell* (1974) and *The Medusa and the Snail* (1979), and occasional writing for the *New England Journal of Medicine* have brought him a large following. *The Youngest Science** will meet his fans' highest expectations.

In American letters I can compare Thomas only with Oliver Wendell Holmes (the father of the one whom Americans think of first). But although both were medical professors and, in their time, deans, their affinity does not really go much deeper than the relaxed and genial style they share: only a young left-wing hothead, insisting always upon relevance and social engagement, would object to being compared with Holmes. As Lewis Thomas is none of these things, he will not mind it being said that his opening paragraph is very much in Holmes's style: 'I have always had a bad memory, as far back as I can remember. It isn't so much that I forget things outright, I forget where I stored them. I need reminders, and when the reminders change, as most of them have changed from my childhood, there goes my memory as well.'

Lewis's father practised medicine in an old clapboard house, the backyard of which abutted on to a cutting that gave way to the Long Island Railroad; and the dilapidated old church which his family attended 'most Sundays' today bears a sign saying that, so far from representing the Dutch Reformed Church, it is Korean Protestant— an apostasy which cannot but strike an Episcopalian as a *non sequitur*. His mother's family were Pecks and Brewsters, who, like everyone

* (Viking, 1983)

else in this part of Connecticut, were rated *Mayflower* progeny—
something which his mother doubted. In the Thomas household
there was never an end to worrying about money—'the family took
it for granted that my father had to worry about his income at the end
of every month'—for few of Dr Thomas's patients paid promptly
and many not at all, though 'some sent in small cheques, once
every few months'. Thus the domestic economy of the Thomas
household could not have been farther removed from that of a great
metropolitan hospital in the USA, with its elaborate security checks
and something akin to a secret police to prevent patients slipping out
without paying their dues.

The practice of medicine was accepted to be a chancy way to make a living,
and nobody expected a doctor to get rich, least of all the doctors themselves
. . . my father's colleagues lived from month to month on whatever cash
their patients provided and did a lot of their work free, not that they wanted
to or felt any conscious sense of charity, but because that was the way it was.

These are considerations to keep in mind; and when, in countries
lacking a national health service, people in need of medical attention
accuse doctors of being mercenary, they would do well to remember
that the profession has a substantial moral credit balance, for there is
no thought or suggestion here that old Dr Thomas was alone or even
unusual in his magnanimity. It was a sign of bad times in the Thomas
household when his mother went foraging round the house for four-
leaf clovers, murmuring: 'The Lord will provide.' This superstitious
ritual had one purpose only: to get old Dr Thomas's patients to pay
their bills. He was not then 'old'—I call him so only to distinguish
père from *fils*.

Lewis Thomas's closest friend as a fourth-year medical student at
Harvard was the Albert Coons who later became one of America's
most distinguished immunologists. It was then Coons's lot to write
the yearbook for the class of 1937. Lewis Thomas was on his
editorial staff and prepared and circulated a questionnaire addressed
to graduates from the years 1927, 1917, and 1907. It dealt with
matters of internship and residency training; and also, delicately and
with a promise of anonymity, asked the respondent for an estimate
of his income for 1937. A space was left for any advice or general

comments that might benefit the class of 1937. 'The average income of the ten-year graduates was around $3,500; $7,500 for the 20-year people,' Thomas reports. 'One man, a urologist, reported an income of $50,000, but he was an anomaly; all the rest made, by the standards of 1937, respectable but very modest sums of money.' The general drift of the advice given by the alumni was that medicine was the best of professions but not a good way to make money: if you could, you should marry a rich wife. In the mean time, work long hours, take very little time off, and don't expect to be prosperous.

Like his fellow practitioners in Flushing, old Dr Thomas worked from home. His consulting-room and waiting-room were on the first floor; the dining room was only one door away from the waiting-room, 'so we grew up eating more quietly and quickly than most families'. The household comprised a living-in maid on the third floor, a laundress who worked in the basement, and a passionate Italian gardener; however, his mother always did the cooking, even when there was a maid, and when there wasn't, she did all the cleaning and everything else in the house.

For those who believe that the riotous misbehaviour of young people is a modern innovation, the outcome of television, non-attendance at Sunday school, and other contrivances of Satan, it is reassuring to know that 'all the children in Flushing were juvenile delinquents, roaming the town in the evening, ringing door bells and hiding, scrawling on the sidewalks with coloured chalk and, at Halloween, breaking windows and throwing garbage cans into front yards.' They shoplifted at Woolworth's, twisted street signs to point in the wrong direction, and at the age of ten were buying and smoking Piedmont cigarettes and breaking candy-vending machines. 'We were expected to be bad, there was no appealing to our better selves because it was assumed that we had no better selves.' An additional cause of grievance was that the children turned out rather well.

In watching his father, Lewis came to learn of the style of medicine that is farthest removed from the modern scientific medicine of which in later years Lewis was to be such an ardent advocate and skilful practitioner. Lewis is full of admiration for his father's unremitting attention and kindliness towards his patients. One

dramatic episode is described of the kind which all laymen and
television producers take for granted as a matter of everyday occurrence in the life of a family doctor. Lewis's father arrived at the
household of an old Flushing family a few minutes after the unmarried
daughter had delivered a baby which the grandmother was about to
smother. His father reasoned with the family and arranged for the
baby to be taken in by the Catholic nuns at the foundling hospital
where the nursing was the best his mother had ever seen. There is, of
course, an element of everybody's father about old Dr Thomas, who
could be heard to swear 'not very demoniacally' when called out of
bed to attend sudden illnesses, births, and deaths, or any one
of a hundred other causes of being scared or in need of help.
Lewis was often taken as a passenger when his father drove off
on his rounds, and was fascinated by the contents of his father's
doctor's suitcase, containing, apart from stethoscope etc., a variety
of ampoules, syringes, and needles. The only really indispensable
drug was morphine; adrenalin was carried in case of an emergency
which never in fact arose: anaphylactic shock. In due course insulin
was added to the pharmacopoeia.

From the standpoint of today's medicine a doctor of his father's
generation had very few if any specific remedies for human ills, but
with the hindsight that Lewis's sympathetic account of his father's
practice makes possible, we can easily see that alongside the good
practitioner's general medical wisdom, kindness, and psychological
support, the older physician had one other factor of immense
importance working in his favour: his patients believed the doctor
and believed that his ministrations would do them good. Indeed,
they believed in medicine as well as in doctors. Nothing else in art
and literature could make this point more clearly than the closing
scene of Puccini's *La Bohème*. Mimi is within a few hours of
death from consumption. Extremities are cold, the pulse barely
perceptible, the breathing light and shallow: but one of the household
resolves to pawn his overcoat and another her trinkets to raise
enough money to buy some . . . medicine! A fat lot of good that
would have done, speaks the voice of modern scientific medicine,
expressing no opinion on the importance of solicitude, both for the
patient and for her anxious friends. The days were yet to come

of a doctor's being quizzed by patients who had read the latest preposterously sanguine article in one or other digest magazine.

Another consideration that worked for the doctor was that patients did get better—some of them, anyway—for 'there are very few illnesses like rabies that kill all comers'. Most illnesses tend to kill some patients and to spare others, 'and if you are one of the lucky ones and have also had at hand a steady, knowledgeable doctor, you become convinced that the doctor saved you'. It is just this element of *post hoc, propter hoc*, I believe, that accounts for the widespread belief that psychoanalytic treatment is efficacious. Of a patient's conviction that the doctor saved him Lewis remarks: 'I should be careful not to believe this of myself if I became a doctor.' The scientific physician now has a lot of fun at the expense of the voluminous and complex prescriptions of previous generations— written out in Latin 'to heighten the mystery'. These medicines differed in taste, colour, smell, and the 'likely effects of the concentrations of alcohol used as solvent'. They had been in use so long that they had the 'incantatory power of religious ritual'. They were prescribed because the patients expected them, not because the physician believed that they did any good except as placebos, and to give the patient something to do while the illness was working its way through its appointed course.

Fads and fashions of therapy had already begun to sweep through medicine in Lewis's father's time. The 'miniscule quantities of drugs' specified by homeopathic principles took hold as a reaction against the powerfully toxic drugs in common use, such as mercury, arsenic, bismuth, strychnine, aconite, and the like. The patients survived as well as they would have done without treatment of any kind and were spared the consequences of intoxication by arsenical and mercurous compounds.

Anyone who has spent any part of their lives in boarding-schools will probably remember how their schooldays were darkened by the capital importance attached to regularity in the movement of the bowels—an attitude of mind still sometimes to be found among old-fashioned nurses and in some cottage hospitals. Lewis Thomas is especially amusing about the theory that represented the very apotheosis of bowel mystique: the theory according to which human disease is caused by the absorption of toxins from the lower intestinal tract—'auto-intoxication'. Thomas's father was persuaded to use on his patients a round lead object about the size of a bowling ball encased in leather. The procedure for promoting peristalsis was to lie the patient flat on his bed and roll the ball clockwise around the abdomen, following the course of the colon. Old Dr Thomas was disappointed in its use and somehow the ball found its way to a neighbour's house, in the garden of which it was rediscovered and announced in the headlines of the local newspaper as a cannon ball from the Revolutionary War, discovered 'to the mystification of visiting historians, who were unable to figure out the trajectory from any of the known engagements of the British or American forces'. Lewis's father, as his son would have done, claimed that he had in a sense made medical history. He was never again caught up in any medical fad. He remained a sceptic about psychosomatic disease. He indulged Lewis's mother by endorsing her administration of cod-liver oil to every member of the family except himself 'and even allowed her to give us something for our nerves called Eskay's Neurophosphates, which arrived as samples from one of the pharmaceutical houses. But he never convinced himself about the value of medicine.' This disenchantment with medicines led him to discover in himself a special talent for surgery, to which he devoted himself exclusively, giving up general practice.

Although he professed a poor opinion of it, Lewis Thomas's memory is in fact an exceptionally good one, a circumstance which gives special strength and point to his historical/sociological comments on the medicine of his own day. His father attended Columbia P&S—the famous College of Physicians and Surgeons of Columbia University—at a time (1901) when medical teaching had already begun to be influenced by the 'therapeutic nihilism' that was the great contribution to medicine of Sir William Osler and his colleagues at Johns Hopkins University. This was the first great scientific revolution in medicine, embodying as it did that element of the scientific ethos which prohibits scientists believing events or phenomena for the existence of which there is no evidence whatsoever. Put in the very crudest terms, obviously in need of much qualification, the teaching of therapeutic nihilism was that medical treatment doesn't do you any good. Indeed, before medical science recognized the nature and importance of hormones, and the pharmaceutical industry made them available in a clinically usable form, and before the sulphonamide drugs and antibiotics added weapons of enormous strength to the physician's armoury, there was not much in the way of medically specific treatment for the physician to use. But during this period the improvement of anaesthesia and the introduction of the aseptic technique made it possible for surgery very greatly to prosper. As late as 1870, a prominent London surgeon had said that the abdomen was 'forever closed to the intrusions of the wise and humane surgeon', and as early as 1900 Berkeley George Moynihan of Leeds was so well satisfied with the progress of surgery as to declare that no further major advances were to be expected. I wonder whether the success of surgery in the early decades of the century, combined with the relative impotence of medicine, was responsible for the traditional rivalry of self-esteem between physicians and surgeons, as I suppose it may have been for the apostasy that made old Dr Thomas give up medicine for surgery.

Therapeutic nihilism, says Lewis, was a

reaction to the kind of medicine taught and practised in the early part of the 19th century, when anything that happened to pop into the doctor's mind was tried out for the treatment of illness. The medical literature of those years makes horrifying reading today: paper after learned paper recounts the benefits of bleeding, cupping, violent purging, the raising of blisters by vesicant ointments, the immersion of the body in either ice water or intolerably hot water, endless lists of botanical extracts cooked up and mixed together under the influence of nothing more than pure whim.

All these things were drilled into the heads of medical students—most of whom learned their trade as apprentices in the offices of older, established doctors. The great revolution brought about by Osler and his colleagues was to have made it clear that most of the so-called remedies in common use were likely to do more harm than good, 'that there were only a small number of genuinely therapeutic drugs—digitalis and morphine the best of all'. The effect of this teaching was that by the time Lewis's father had got to P&S the centre of gravity of medical treatment had shifted to diagnosis—the recognition of specific illnesses, if possible through an understanding of the natural history of disease.

Under the chapter heading '1911 Medicine' Lewis Thomas writes with Hippocratic gravity and authority on the desiderata in a graduate of one of the great medical schools after the Oslerian revolution:

Most medical students of those decades had hard things to learn about. Prescriptions were an expected ritual laid on as a kind of background music for the real work of the 16-hour day. First of all, the physician was expected to walk in and take over; he became responsible for the outcome whether he could affect it or not. Second, it was assumed that he would *stand by*, on call, until it was over. Third, and this was probably the most important of his duties, he would explain what had happened and what was likely to happen. All three duties required experience to be done well. The first two needed a mixture of intense curiosity about people in general and an inborn capacity for affection, hard to come by but indispensable for a good doctor. The third, the art of prediction, needed education, and was the sole contribution of the medical school; good medical schools produced doctors who could make an accurate diagnosis and knew enough of the details of the natural history of disease to be able to make a reliable prognosis. This was all there was to science in medicine, and the store of information which made diagnosis and prognosis possible for my father's generation was something quite new in the early part of the 20th century.

To this we must add that the physician required a close knowledge of what good nurses were able to do (Lewis's mother had trained as a nurse): 'The nurses had their own profession, their own schools and their own secrets.'

Lewis's father is the protagonist of the chapter '1911 Medicine': Lewis himself steps into the limelight in '1933 Medicine'. There was not then today's fierce competition to enter medical school, and although Lewis's record at Princeton was only 'middling fair', he was received into Harvard, perhaps (in the usage of the day) with a bit of help from his parents' friend, the admirable Dr Hans Zinsser, who was later to write *Rats, Lice and History*. Though Lewis's education was not unlike his father's in principle, a great flowering of medical science in physiology and biochemistry, assisted by microbiology and immunology, had taken place since his father's time and 'transformed our understanding of the causation of major infectious diseases. But the *purpose* of the curriculum was, if anything, even more conservative than thirty years earlier.' Its purpose was to teach the classification and recognition of disease entities, now in their laboratory as well as their clinical manifestations: 'the treatment of disease was the most minor part of the curriculum, almost left out altogether.' Pharmacology had to do with the mode of action of a handful of useful everyday drugs, such as aspirin (the discovery of the mode of action of which was the work of the past few years), morphine, barbiturates, and digitalis, plus a few others. Lewis could not remember any occasion upon which his instructors referred to the contents of the thin book called *Useful Drugs*. Indeed, he did not 'remember much talk about treating disease at any time in the four years of medical school except by the surgeons.'

Their own classmates were probably the most important influence on Harvard medical students in Lewis's day, and it was through their anxious colloquies that third- and fourth-year students came to realize that they didn't know much that was really useful and that 'we could do nothing to change the course of the great majority of the diseases we were so busy analysing, that medicine, for all its façade as a learned profession, was in real life a profoundly ignorant occupation'.

Happily any danger of disenchantment with medicine was obviated in his fourth year when, during his clinical clerkship at Beth Israel, he watched with delighted admiration the conduct of a complete physical examination by Professor Blumgart. Blumgart was an enormously perceptive physician with great intuitive powers—by which I understand a man who finds his way to a conclusion along logically unscripted pathways. Moreover, Lewis remarks that, so far as he knew, Blumgart was never wrong, not once. But he can recall 'only three or four patients for whom the diagnosis resulted in the possibility of doing something to change the course of the illness, and each of these involved calling in the surgeons to do the something . . . For the majority, the disease had to be left to run its own course, for better or worse.' This being so, it is easy to see how there quite quickly grew up, and why there still persists, a sort of tension between physicians and surgeons, especially among people who are not very good at either medicine or surgery. Physicians have long thought themselves the patricians of medical practice, choice spirits altogether more intellectual and refined than those bustling, sanguine, adventurous, and self-confident surgeons who often did more for their patients than the physicians could.

Two of the five principal services of the Boston City Hospital, 'the city's largest, committed to the care of indigent Bostonians', were staffed by Harvard Medical School, and when Dr Francis Weld Peabody, reputedly the best of its physicians, founded a clinical research centre with the Thorndike Memorial Laboratories on the hospital campus, the hospital became an irresistible attraction to the brightest physicians. By the time Lewis Thomas got there in 1937, the staff was formidably talented, their names still remembered for their contribution to the causal interpretation and (by means of liver extracts) treatment of pernicious anaemia. The Thorndike laboratory became a model for establishing a working liaison between laboratory and ward.

It was in the mid-1930s that the great gun-turret of scientific medicine swung round to train upon infectious disease. If treatment could be started early enough in the course of the disease, Ehrlich's neoarsphenamine was effective in the treatment of syphilis. The great leap forward, however, was the development out of the dyestuff *prontosil* of the anti-bacterial substance paraaminobenzene-sulphonamide (sulphanilamide). This was the founder member of the great sulphonamide family of drugs, the oral administration of which would control puerperal sepsis and streptococcal meningitis. In later years, the triumphant success of these synthetic bacterials had the odd and unexpected effect of retarding the development of penicillin, because when money was sought to promote research into such antibiotics, wise guys on grant-giving committees announced that the future of antibacterials lay with synthetic organic chemicals such as Gerhard Domagk had introduced and not with obscure medieval-sounding nostrums like extracts of fungi.

In this period, so Lewis tells us, 'immunology was beginning to become an applied science'. This took me aback: for immunology, though not then so described, scored its first triumph with the success of *variolation*—that is, deliberate infection with a pustule from a mild case of smallpox in order to protect against the virulent disease—

but the first consciously immunological procedure was the raising of antibodies against pneumococci using the purified polysaccharide capsules of the bacteria to stimulate antibody formation.

Antibodies were not then an unqualified success—and never became so, the trouble being that the vehicle of the antibody is necessarily a blood serum such as horse serum, which can arouse overwhelming and sometimes fatal anaphylactic reactions in those into whom they are injected. Lewis recounts many of the medical triumphs that preceded or accompanied the period of his internship at the Boston City Hospital: pellagra became curable; insulin had been isolated by Banting and Best; treatment with liver extract removed pernicious anaemia from the roll of uniformly fatal diseases. Lewis makes it all sound, as indeed it was, a great time for a medical scientist to be alive, for here were modern secular miracles: sulpha-nilamide, for example, snatched from what would otherwise have been certain death patients already moribund from pneumococcal and streptococcal septicaemia. For an intern, it was like the opening of a new world: 'We had been raised to be ready for one kind of profession, and we sensed that the profession itself had changed at the moment of our entry.'

Lewis thought of his internship quite simply as 'the best of times', despite the cruelly hard work, the long hours and the meagre pay. This last was not a grave embarrassment. None of the interns was married and there was no time to spend money. Two blood donations a month at $25 a time 'kept us in affluence'—and in liquour too, for a state law stipulated that a blood donor was entitled to a pint of whisky. The most junior intern ('the pup') spent his 24-hour day collecting specimens of blood, urine, faeces, spinal fluid, and sputum, and doing all the laboratory diagnostic work. Two further trimesters, each with special responsibilities, completed the internship; the next rung on the ladder was Assistant House Physician, and then, after 15 months' service, House Physician.

I was pleased to read in Lewis Thomas an account of something a layman is likely to learn about only from fiction: the nature of 'one of the great phenomena of human disease—the *crisis*', as it might occur in a young adult victim of lobar pneumonia. Typically, the disease ran ten to fourteen days with a high fever each day, the patient

suffering more and more chest pain and getting more and more
alarmingly exhausted, and nearer and nearer the shores of Lethe. If
the specific antibodies made by or injected into the patient prevailed
over the pneumococci, then one day the patient's temperature
would plummet from 106 degrees to normal while he sweated
copiously. He would then announce that he felt better and would
like something to eat.

A second great emergency was diabetic coma. If the diagnosis was
right, and the appropriate treatment given, recovery was certain. All
possible help was sought and 'the senior visiting physicians came
across the ramp from the Thorndike on the run'. Then there was
acute heart failure. There were only three ways of coping with it,
'not always effective and never in any sense curative'. The first was
to withdraw a pint of blood from an arm vein to reduce the
hydrostatic burden on the heart; the second was to inject digitalis—
a nice matter, for too little would be ineffective and too much
toxic; the third expedient was breathing oxygen. Syphilis, even if
recognized early enough in its course, made an intern's heart sink
at the prospect of months or years of administering arsenicals,
mercurous compounds, and bismuth with all the attendant risks of
toxic side-effects. When the spirochaetes reached the brain nothing
could be done. Although syphilis is again on the increase, it is
now manageable because of the susceptibility of the spirochaete to
penicillin.

Alcoholics were sufficiently numerous for the City Hospital to set
up a 40-bed ward just for them. Delirium tremens was as bad a sight
as anything that came an intern's way. The treatment was a huge
dose of paraldehyde, vitamin B, liver extract, and ice packs to reduce
the fever: 'That was it for the DTs, and we saw a lot of deaths.' One
death at the City Hospital was the outcome of what Lewis describes
as the worst mistake he ever witnessed. A young black musician
was admitted shivery and apathetic—ostensibly a victim of the
pneumonia just then in season. In the event, malignant malaria was
diagnosed—but too late for the administration of quinine, which
might have saved him. How on earth did he get malaria, having
regard to the fact that the anopheline mosquito shuns Boston in the
winter? The young musician was a heroin addict, accustomed to

parties at which the needle was passed from hand to hand. The medical staff felt guilty and ashamed, as William Osler said they should in the first sentence of the chapter of his textbook devoted to malaria. 'It was a bad day for Harvard.'

Lewis's thoughts were not then and are not now wholly preoccupied by the day-to-day practice of contemporary medicine: his mind restlessly seeks philosophic or philological problems to engage in. 'What did doctors *do*?' Lewis asks, reflecting that plague, typhus, tuberculosis, and syphilis were representative of those many infections whose progress was rapid and whose outcome was usually lethal. For one thing, they practised a little magic, 'dancing around the bedside, making smoke, chanting incomprehensibilities and *touching* the patient everywhere. This touching was the real professional secret, never acknowledged as the central, essential skill.' Lewis rates touch as the oldest and most effective act of healing. 'Some people don't like being handled by others, but not, or almost never, sick people': 'part of the dismay in being very sick is the lack of close human contact'. In the course of time, touching, like everything else in medicine, became more specialized and refined, and turned into 'palpation'—feeling for the tip of the spleen, or the edge of the liver—or into a thumping of the chest in order to ascertain whether the sound was dull or resonant. The gift possessed by these doctors who began the laying-on of hands was probably the gift of affection. Certainly people who do not like other people very much would have been likely to stay away from an occupation that required touching. Touching reached its highest point, Lewis thinks, in the practice of laying the naked ear against the front or back of the chest, adopted as soon as it came to be known that heart sounds could be very informative. However, the tendency of technology to increase the distance between doctor and patient, which began with the introduction of the stethoscope in the nineteenth century, has now gone so far that a physician may remain in his office while his patient is in another building. Physical contact is confined to a perfunctory handshake: 'medicine is no longer the laying-on of hands, it is more like the reading of signals from machines'. The mechanization of scientific medicine is here to stay, but the patient may well feel that the doctor is more interested in his disease than in himself as a person. In his father's time, Lewis remarks, talking with the patient was the biggest part of medicine, for it was almost all there was to do.

Old Dr Thomas's professional observations on the work of nurses, combined with his good sense in having married one, gave him 'a deep and lasting respect for the whole nursing profession'. As an experienced, longtime, virtually award-winning hospital patient myself, I can only agree. In Florence Nightingale's day, nurses were looked down upon as loose women and this may account for the hospital tradition that nurses must be kept occupied the whole time: in the unlikely event of their being momentarily unoccupied, work must be found for them. This is fully in keeping with Dr Thomas Arnold's teaching on the way to treat public schoolboys. The reverend doctor's message to mankind was that if schoolboys were unoccupied for more than two minutes they would inevitably bugger each other. I share with Sir John Betjeman the fate of having been educated at a public school so steeped in the Arnold tradition that, when we were there, water closets were doorless: but perhaps the Laureate was exaggerating when, later in life, he said that he did not have a bowel movement for three years.

I went into a big London teaching hospital for a minor operation about 1935 and was horrified to see the way work was found for nurses: hospital bandages and dressings were packed into cylindrical metal drums for sterilization by superheated steam. These drums were made of nickel-plated copper and it was a rule that they should be kept dazzlingly polished at all times, so that already overworked, pale, and tired-looking nurses could be seen in the ward at all times with little pots of metal polish, rubbing, rubbing and rubbing away. The spirit of Dr Arnold still glared so brightly out of the eyes of the senior nurses that they would have judged it bad for discipline if all that polishing had been done away with, if, for example, the obvious solution of using oxidized metal sterilizer drums had been adopted.

I was specially pleased to see that Lewis Thomas devotes a chapter to neurology, since this is the field in which, as a hospital patient, I specialized. Neurology today is very like medicine in general fifty to a hundred years ago, in its preoccupation with interpretation and diagnosis and the relative backwardness of treatment. But in

Lewis Thomas's day, neurology was undergoing a transformation: penicillin and the antibiotics generally have made brain abscesses less of a threat than hitherto and the pharmacologists have devised effective anti-convulsant drugs. Multiple sclerosis is still a most terrible evil, though long before such thoughts became fashionable, Lewis Thomas formed the opinion that multiple sclerosis was an auto-immune disease. This is still the best aetiological bet upon which new schemes of treatment are constantly being devised and tried out. Meanwhile myasthenia gravis, another form of creeping paralysis, has been shown quite definitely to be auto-immune in character and, as such, therapeutically manageable. In addition, the coming of cryosurgery (the destruction of tissue by exposure to temperatures reaching 200 degrees below zero Celsius) and of computer-assisted tomography (CAT) for finding out what is wrong inside has added greatly to the skills of the neurosurgeon. These, and later on the recognition of our own endogenous analgesic drugs, the endorphins, have made neurology for Lewis Thomas the most fascinating field of medicine. Lewis is authoritative on the subject of cancer, in a chapter devoted to which he reaffirms his belief and hope that the notion of immunologic surveillance will turn out to be valid. He also believes that cancer is not a generic name for a whole variety of different diseases, each of which has a different aetiology and requires a different treatment; cancers have enough in common to justify their being thought of as an aetiological entity.

Lewis Thomas held an unexpectedly large number of administrative appointments: he has been departmental head, dean, president, and chancellor. I say 'unexpected' because Lewis Thomas derives no pleasure from the exercise of power and has never had need to advance his career by such onerous and time-consuming means. Only duty, then, can ever have been his motive for taking administrative office and one cannot but honour him for doing so. Writing on 'the governance of a university', he is at his sparkling best:

How should a university be run? Who is really in charge, holding the power? The proper answer is, of course, nobody. I know of one or two colleges and universities that have actually been tightly administered, managed rather like large businesses, controlled in every detail by a president and his immediately surrounding bureaucrats, but these were not really very good colleges or universities to begin with, and they were managed this way because they were on the verge of running out of money. In normal times, with institutions that are relatively stable in their endowments and incomes, nobody is really in charge.

It is natural to ask what kind of science we should expect from a man who writes and thinks as Thomas does. The answer is something exceptionally perceptive and witty, and unexpected in its juxtaposition of ideas. His first field of interest is that of immunological surveillance, as popularized by the writing of Sir Frank MacFarlane Burnet, an idea so clarifying that almost everyone in the business wants it to be valid. The idea can be said to grow out of a consideration of the teleology of the process by which foreign (that is, 'non-self') organ tissue grafts are recognized as such and rejected by an immunologic process. Lewis Thomas formed the opinion that the rejection of grafts was a tiresome by-product of the existence in the body of a monitoring system of which the primary purpose was to spy out and identify non-self variants among the cells in the body that might give rise to malignant growths which could then be reacted upon by lymphocytes much as foreign grafts are acted upon and rejected. This is a lovely idea and it ought to be true, but it is not universally accepted: there are even doubts today as to whether tumours are normally destroyed by a *bona fide* immunologic process— doubts that can only be resolved by an *experimentum crucis* which, though I believe it to be experimentally feasible, has not yet been performed. Immunological surveillance has none the less been an enormously fertile theory, by means of which we have learned a great deal more about the nature and behaviour of tumours than we should otherwise have known.

The second of my favourites among Lewis Thomas's ideas could also be deemed to have a teleological origin. It is well known that living tissues cannot normally be transplanted from one human being to another, or from one mouse to another, or from one goldfish to another. This is because all these animals differ from others of their kind in the make-up of the genes that control transplantability. These differences are inherited and the genes that control them form a so-called polymorphic system (a stable subdivision of the population into different genetic types that persist from generation to generation in roughly the same proportions). This is true of blood groups, too.

What is the point of this polymorphism and what keeps it going? This is a question that must be asked of every polymorphism, and is often answered. In collaboration with a cancer research worker from his own diocese, the Sloan-Kettering Institute, Dr Edward Boyse, a man cast in much the same mould as himself, Thomas propounded the idea that these ingrained differences in tissue transplantation groups determine mating preferences.

This is something that could only be mediated through smell: mice which have different make-ups with respect to genes affecting transplantation must smell different from each other. If this difference in smell does indeed exist, the reasoning went, a dog must be able to tell one mouse from another, even if the mice differ by only a single gene. Put in this way, the hypothesis was clearly testable. I watched an enormously informative test in progress at the Medical Research Council's Clinical Research Centre in London—happily situated near the Hendon Police College, widely regarded as the world's greatest centre for the training of tracker dogs. What I found so deeply instructive was the system by which the tracker dogs were trained. Human mothers and pedagogues are reconciled to the notion that training is the outcome of a judicious blend of rewards and punishments. Unless I was being unduly inattentive, these tracker dogs were trained only by rewards, which took the following form. When the tracker was given some nesting material from a mouse's cage so as to get the mouse's smell into its mind, it was then set the task of picking this mouse out from a number of others. If it did so, it would bound back to its trainer, who thereupon patted the dog enthusiastically, scratching its tummy and exclaiming inanely: 'There's a good dog!' (Pat, pat, tickle, tickle.) 'Who's a good dog, eh?' (Pat, pat.) '*There*'s a good dog.' (Tickle, tickle.) And so on. I did not take part in the execution of this experiment: I merely 'assisted at the experience', as our French colleagues say. I still wonder what lesson we can learn from the apparent truth that, for these tracker dogs, the withholding of approval is punishment enough.

14

The Meaning of Silence
(1984)

Lewis Thomas's latest book* is a collection of 24 short essays of which the first has to do with the gravest problem confronting mankind—the Bomb. In this essay his fans see a different Lewis Thomas, angry where he was once urbane, grim rather than gay, for no aspect of the bomb is at all funny and upon this subject Thomas is unrelievedly grave. His night thoughts are akin to those that most of us have when awake at dawn or sleepless in the small hours of the morning, or whenever the faculty of self-deception that so often insulates us from real life is temporarily in abeyance. For me, the gravest of these black morning thoughts is that the future of England and, ecumenically speaking, of the world depends upon the decisions of party politicians such as Mr Heseltine† who can have no deep understanding of these awful matters, and of warlords who in respect of strategic understanding and common humanity are not likely to have altered greatly from, for example, Field Marshal Douglas Haig, architect of the strategy of attrition that cost hundreds of thousands of British casualties in the Somme offensive and at Passchendaele.

Lewis Thomas considers first the failure of military planners clearly to envisage the 'worst case' among the possibilities they profess to consider. 'At the outset of World War One the British didn't have in mind the outright loss of an entire generation of their best youth, nor did any of the Europeans count on such an unhingeing of German society as would lead straight to Hitler.' Again, 'defeat at the end was not anywhere on the United States' list

---

* *Late Night Thoughts on Listening to Mahler's Ninth Symphony* (Viking, 1984).
† The then British Secretary of State for Defence.

of possible outcomes of the Vietnam adventure—nor was what happened later in Cambodia and Laos part of the scenario.'

'The final worst case for all of us has now become the destruction by ourselves of our species.' This would not be a novel event biologically: several species have gone under before in the history of life on earth. Where now are the trilobites and the great reptiles of the mesozoic? However, Thomas thinks that the environmental changes or epidemics that did them in could never endanger a species as intelligent and resourceful as ours: 'We will not be wiped off the face of the earth by hard times, no matter how hard: we are tough and resilient animals and are good at hard times. If we are to be done in, we will do it ourselves by warfare and thermonuclear weaponry and it will happen because the military planners and the governments who pay attention to them are guessing at the wrong worst case.'

There follows a paragraph written very much as Voltaire would have written it if he had compiled his *Dictionnaire Philosophique* today:

Each side is guessing that the other side will, sooner or later, fire first. To guard against this, each side is hell-bent on achieving a weapons technology capable of two objectives: to prevent the other from firing first by having enough missiles to destroy the first-strike salvo before it is launched (which means, of course, its own first strike) and, as a back-up, to have for retaliation a powerful enough reserve to inflict what is called 'unacceptable damage' on the other side's people. In today's urban world, this means the cities. The policy revision designated as Presidential Directive 59, issued by the Carter White House in August 1980, stipulates that enemy command and control networks and military bases would become the primary targets in a 'prolonged, limited' nuclear war. Even so, *some* cities and towns would inevitably be blown away, then doubtless more, then perhaps all.

In reading Thomas's account of the acute (i.e. direct, proximal) effects of the Hiroshima bomb we *must* keep it in mind that by modern standards this was a pipsqueak affair—a 'technological antique' like a Tiffany lamp. With a modern thermonuclear bomb nothing would remain alive within a radius of six miles from the hypocentre of the explosion—the 'hypocentre' being the point on the earth perpendicularly beneath the exploding bomb—and it is a most mischievous illusion that politicians, company presidents, ministers, and other high-ups, having gone to ground during an explosion of which the precise time and place are assumed to be known beforehand, could now emerge to re-establish order and communications and institute medical treatment. Of the latter, Lewis Thomas remarks that a person exposed to near-lethal irradiation *can* occasionally be saved, though only by using the full resources of a highly specialized hospital unit, with endless transfusions and bone marrow transplants: so what could be done for a thousand such cases all at once, or a hundred thousand—quite apart from the multitude of conventionally maimed or burned people? The words 'disaster' and 'catastrophe' are too frivolous, says Thomas, to describe the events that would inevitably follow a war with thermonuclear weapons.

Lewis Thomas quotes extensively from *Hiroshima and Nagasaki: The Physical, Medical and Social Effects of the Atomic Bombings* and notes that Hiroshima was spared assault by conventional war weapons 'in order to measure with exactitude the effects of the new bomb'. He also remarks that when American journalists reported on the effects of the bomb about a month after detonation they described only the physical damage, and no news about injuries to the people, especially news about radiation sickness, was allowed by the Allied Occupation. Moreover 'on 6 September 1945, the General Headquarters of the Occupational Forces issued a statement that made it clear that people likely to die from A-bomb afflictions should be left to die. The official attitude . . . was that people suffering from radiation injuries were not worth saving.'

In a long-term view, it is not the acute but the chronic effects of thermonuclear weaponry that are the greater threat to mankind: the cancers, for example, that occur with higher than normal frequencies for decades after a nuclear explosion, and the shortening of the life-span which has been observed in experimental situations to be among the long-term consequences of sub-lethal irradiation, not to speak of the chromosomal damage and the genetic consequences in general. All is darkness and Lewis Thomas now gives us a new menace to think about:

a good-sized nuclear bomb, say ten megatons, exploded at a very high altitude, 250 miles or so over a country, or a set of such bombs over a continent, might elicit such a surge of electromagnetic energy in the underlying atmosphere that all electronic devices on the earth below would be put out of commission—or destroyed outright—all computers, radios, telephones, television, all electric grids, all communications beyond the reach of a human shout. None of the buttons pressed in Moscow or Washington, if either lay beneath the rays, would function. The silos would not open on command, or fire their missiles. During this period the affected country would be, in effect, anaethetised, and the follow-on missiles from the other side could pick off their targets like fruit from a tree. Only the submarine forces, roaming far at sea, would be able to fire back, and their only signal to fire would have to be the total absence of any signal from home. The fate of the aggressor's own cities would then lie at the fingertips of individual submarine commanders, out of touch with the rest of the world, forced to read the meaning of silence.

How should ordinary people react to the prospect of these calamities? In my view, the avoidance of nuclear warfare is a consideration of overriding priority—something more important than national sovereignty, 'face', or prestige, or, of course, the continuance in office of party politicians. We must resolutely abstain from any political decision that might promote or be thought to promote our candidature as a thermonuclear target, and we must be seen to be incapable of mounting a first strike on our own account or as an agent of NATO. If these are unconditional principles, I see no alternative to unilateral nuclear disarmament. The advantage that would be taken of such an action by a potential enemy is merely suppositious, but the effects of thermonuclear assault are for real. Thomas reminds us that the words REST IN PEACE FOR THE MISTAKE WILL NOT BE REPEATED are carved in the stone of the cenotaph at Hiroshima, but 'meanwhile the preparations go on, the dreamlike rituals are rehearsed, and the whole earth is being set up as an altar for a burnt offering, a monstrous human sacrifice to an imagined god with averted eyes'. If our rulers are not, as they sometimes seem to be, unteachable, what lesson could be more clamant or more brutally exigent than that which has been taught by Hiroshima and Nagasaki? What can we peaceable people teach them that Hiroshima did not?

Not all of Lewis Thomas's book is in the dark and foreboding vein of this first chapter; the witty and urbane New Yorker we are more familiar with wrote the remaining chapters. It is praise enough to describe them as vintage Thomas—full of good things, unexpected aperçus and witty juxtapositions of ideas. An important element of Thomas's style is to be reassuringly dismissive about the imagined threat of exaggeratedly scary things such as artificial intelligences. Although computers can manufacture successions of sounds 'with a disarming resemblance to real music', he says, language is a different matter. There would be no problem, he says, in constructing a vocabulary of 'etymons'—an excellent word for a purely nounal vocabulary of simple designations: 'the impossibility would come in providing for the ambiguities, metaphors and mistakes characteristic of real language.' And what computer, I wonder, could differentiate reliably between *nonne* questions and *num* questions: those that expect the answers yes and no respectively.

Thomas loves words and, taking an educated readership for granted, he assumes that his readers do too, so his books always contain a number of philological divertissements: thus he tells us that 'the word "gibberish" is thought by some to refer back to Jabir ibn Hayyan, an eighth-century alchemist, who lived in fear of being executed for black magic and worded his doctrines so obscurely that almost no one knew what he was talking about'. One of Thomas's most attractive gifts is for comical hyperbole: 'a family was once given a talking crow for Christmas, and this animal imitated every nearby sound with such accuracy that the household was kept constantly on the fly, answering doors and telephones, oiling hinges, looking out of the window for falling bodies, glancing into empty bathrooms for the sources of flushing water.' He loves the English language and deservedly exults in his ability to write it so well. If he had not been a medical scientist he would have been happiest, I think, as a philologist: one of his essays, indeed, is on the birth and growth of a new language, Hawaiian Creole, in a polyglot community that had formerly relied upon pidgin English—'pidgin', I learned, being the pidgin for 'business'.

Later on Lewis Thomas's thoughts return to the Bomb and the sky darkens again: for by simple reasoning to do with the immense difficulty, complexity, and expense of the procedures used to rescue the victims of burning and radiation injury, and with the rarity of the people qualified to put them into effect, he infers that modern medicine has nothing whatever to offer, not even a token benefit, in the event of a thermonuclear war. This is a very grave statement and no one in the world is better qualified than Thomas to make it: he is thoroughly familiar with and has contributed to the science underlying the medicine and surgery of repair and has a thorough understanding of the administration and execution of practical medicine. There is, incidentally, no foundation whatsoever for the idea that seems to underlie much of Britain's strategic thinking on the bomb: that if we are the target of a thermonuclear attack, British grit and the Dunkirk spirit will see us through and—more than that—will unite us all in warm comradeship and resolution for victory, as the bombing did in the Second World War. Thomas plaintively asks what has gone wrong in the minds of statesmen in this generation, and how it should be possible that so many people with the outward appearance of steadiness and authority, intelligent and convincing enough to have reached the highest positions in the governments of the world, should have lost their sense of responsibility for the human beings to whom they are accountable? It is to psychiatrists and social scientists that he looks for an answer. What a hope! The case is much too serious for the glib psychologisms of psychiatrists and the lame fumblings of social science—and in any case, Thomas remarks, if we are going under it would be small comfort to understand 'how it happened to happen'.

The problem is not insoluble in the sense that it is mathematically impossible to devise a straight-edge and compass construction to trisect an angle or to 'square the circle'—that is, to draw a square equal in area to a given circle—or that it is logically impossible to arrive at transcendental theorems from the axioms and observation statements of science, containing, as they do, only empirical furniture. No: the problem is not insoluble, but it is too difficult—too far beyond the capabilities of warlords and politicians whose judgments of priority are obfuscated by considerations of political or economic advantage, 'face', and national prestige. Because of his bluntness and strength of mind and great personal authority, Lewis Thomas has performed an important public service—and with the most enviable literary grace.

15

The Cost-Benefit Analysis of Pure Research
(1973)

It is often said by people who sound as if they thought they were saying something original that ideas are the lifeblood of research and that money cannot buy ideas. Administrators of such lowly intellectual accomplishments that they regard this excruciating platitude as the embodiment of an original thought are sometimes tempted to infer that because money is not *sufficient* to advance research it is not *necessary*; and they take very naturally to that bygone romantic literature of research of which the principal message was any *really* first-rate scientist could get away with apparatus no more elaborate than a student's microscope and two or three empty Heinz bean cans. This is a pernicious doctrine because the truth is so utterly different. I learned from my own early days in research that if one lacks adequate equipment—e.g., an ultracentrifuge or facilities for radioactive labelling and counting—then some internal censorship of unknown circuitry prevents one having ideas of the kind that could only be evaluated by means of such equipment. Money can't buy ideas, that's for sure, but lack of it can prevent one having them.

It is unhappily in the field of cancer research generally that the ideas/money platitude is heard most often: it is often solemnly affirmed that what is most needed in cancer research is more ideas, not more money. There couldn't be a more unfortunate example. Anybody who knows the great cancer research institutes from the inside—e.g., the Sloan-Kettering in New York and the Chester-Beatty in London—knows that there never was a time when research workers had so many bright and promising ideas. I very strongly suspect that when some general cancer therapy is devised it will be built upon the principles and ideas that are already commonplace in the discussions and research agenda of cancer research workers in the institutes I have mentioned. My personal view is that the principal need of cancer research just at the present is to devise experimental models—very likely based on the use of primates—which will act as a go-between to bridge the worlds of mice and men. Moreover, more posts must be created for people with one foot in the ward and one in the laboratory: people who can bring possible therapeutic procedures to the bedside more quickly than now seems to be possible and who can supervise or, if necessary, execute clinical trials themselves. Anybody who thinks that all this can be done without the expenditure of very large sums of money should divulge the secret of *how* to do so forthwith: we are all agog.

But oncology does raise in a specially acute form the problem of assessing the cost-effectiveness of research, for everybody knows that extremely large sums of money have *already* been awarded to cancer research and knows also that what laymen injudiciously call *the* cancer problem has not yet been solved. *It will be solved*, however, and when it is the large sums spent on it will be retrospectively justified and misty-eyed authors of Commencement Addresses will be talking about a 'priceless' or 'invaluable' benefaction to mankind, although at the same time, without any awareness of inconsistency, they may well put on solemn and businesslike expressions and call for a realistic pricing or valuation of the benefits of research to mankind.

However, GNP, the tribal God of the western world, can be propitiated only by a prospective, not merely a retrospective, justification of the money spent on research. The problem is to determine now what the future return will be on, say, $1 million spent on biomedical research today.

The problem of determining the cost-effectiveness of basic research need not be a soluble one and I submit that it is not. Indeed, I believe it to be *essentially* insoluble. I have already explained elsewhere* that the notion of predicting future ideas or future theories involves a

* In *The Art of the Soluble* (Methuen, 1967).

logical self-contradiction. It simply can't be done. It is therefore not possible to predict the pathway of future research. In *The Open Society and Its Enemies*, Karl Popper, whom many regard as the leading philosopher of the western world, long ago made it clear that similar considerations apply to the social sciences. A single example must suffice to make his point clear: with steam as the source of power it must be specially economical to energize machine tools by means of flexible belts connected with a single rotating shaft that can serve a great many tools and production processes under one roof. This being so, Karl Marx was quite right to infer that means of production must become ever more concentrated around the source of power, so that the cottage industry must eventually be supplanted altogether by the 'satanic mill'. But of course Marx could not foresee the advent of the electrical power which has made possible the innumerable small light-engineering and electronics factories (e.g., in Massachusetts or the English midlands) which provide a modern, scaled-up equivalent of the cottage industry.

Although it may, not unjustly, have the character of a *reductio ad absurdum* it is worth while outlining the kind of information we should need to feed into a computer in order to have any expectation of getting an answer to the problem of costing the benefits of research. We need to know the urgency of the problem assessed in terms of the cost of *not* solving it. We need at the same time to evaluate precisely the urgency of competing claims on the limited sums of money available. Then again, if the research professes, say, to find a remedy for the dread disease of infectious omphalitis we shall somehow have to ascertain the contribution to society, positive or negative, that would have been made by any future victims of the disease. Maybe the nation would be better off morally and moneywise if some of the potential beneficiaries of the research did in fact succumb to the disease it was designed to prevent; maybe not—who knows? Nobody knows, of course, and what is more, nobody *can* know. Even when we have attached numerical values to all these variable quantities, some of which can hardly be so expressed, we still have to devise a multivariate function in all these variables and also in $t = $ time, the solution of which will enable us to predict the future return on present research expenditure.

The accusation is sometimes directed against scientists that there is in reality no such thing as *the* scientific method, i.e., that there is no logically accountable and intellectually rigorous process by which we may proceed directly to the solution of a given problem. Scientific method works only in retrospect. This accusation is perfectly just but it doesn't in practice amount to anything more than saying that there is no set of cut-and-dried rules for writing a poem or passage of music or conducting any other imaginative exercise. Another variant is to speak contemptuously of scientists' playing 'scientific roulette', a jealous accusation which in my experience is most often made by scientific hacks who have spent many laborious years on inductive fact-hunting exercises without even interesting anybody very much. The contemptuous reference to 'scientific roulette' is a method of cutting their more successful and probably more gifted colleagues down to size.

I choose now an example borrowed shamelessly from a learned and witty discourse by one of England's leading physicians (Sir John McMichael) in introducing a lecture series on 'The Scientific Basis of Medicine'. It illustrates both the unpredictability of ideas and the impossibility of costing the long-term consequences of scientific research. Let us suppose that a funding and research support system similar to today's prevailed in the 1890s and that a grant-giving body received a research proposal to make the human body transparent, so as to enable orthopaedic surgeons to visualize bones and joints directly through the flesh. It takes no great feat of the imagination to picture the slow regretful shaking of wise grey heads as they pronounce the enterprise impracticable. Yet we all know that the enterprise came off and that its medical possibilities were implicit in the very earliest reports of it, but even in retrospect—knowing everything we have learned since 1896—it would be impossible to fix a cash value to the discovery. The existence of radiography and radiotherapy as a facility has been an added public expense in those countries which have a national health service; we can hardly expect to cost the loss to society of those unfortunate pioneers who lost their lives through exposure to x-radiation before its malignant properties were understood. I am afraid we shall have to regard the funding of 'pure' research as a tax levied upon society that is not dissimilar in kind from that which maintains art galleries and opera houses—a 'civilization tax', perhaps.

The cynical attitude towards the pretensions and demands of 'pure' research that is now fashionable among economists and others who would like to pass for realists might have sprung into being in order to provide a context for Oscar Wilde's definition of a cynic: 'A cynic', said Wilde, 'is a man who knows the price of everything and the value of nothing.'

16

The Pure Science
(1973)

In the bustling mercantile world which gave birth to the first great professional scientific society—The Royal Society of London—its two leading spokesmen, Sir Francis Bacon and Thomas Sprat, had to plead that not all scientific research should be done for immediate practical use. In addition to 'experiments of use' there were to be 'experiments of light'. This was the first plea that pure science should not be submerged by the insistent and urgent demands of practical application.

Bacon himself was quite clear that applied science itself could not flourish except in the light that was to be thrown by the luminary experiments which the first members of the Royal Society were exhorted to perform. Since then the balance between pure and applied science has swayed to and from under the weight of public opinion.

After the war it was very necessary to insist again upon the merits of the life of the mind and of a tranquil pursuit of truth. Today, however, in their anxiety to propitiate GNP, the tribal god of the Western world, whose name (like that of JHVH) the very pious may refer to by consonants only without an irreverent vocalization of the intervening vowels, people are tending again to sneer at pure learning and are beginning to ask themselves if a university or any other agency should subsidize research upon the lesser-known laundry bills of D. H. Lawrence or on the nature of the black spots that sometimes appear on some eggs of some sea-urchins.

Yet in science and technology it is the hard-faced, practical-minded, no-nonsense businessman or man of action who over and again turns out to be an impractical dreamer with his head in the clouds: the simple unworldly fellow who believes that one can reap a crop of practically useful and financially rewarding research results without a long and careful preparation of the soil.

Lord Rothschild, himself a member of the Royal Society and also a chairman of the Government's so-called think tank, has caused quite a flurry in the world of learning by recommending legislation based on the Government's consumer/contractor principle. The consumer (often a department of Government) orders what he wants in the way of results and pays the contractor if he gets them, or thinks he is likely to. It is a grave weakness of Rothschild's recommendations that they do not provide adequately for the prosecution of pure research. Some of his public utterances have been marred by embarrassing vulgarisms, as when he dismissed the doctrine of the Unity of Science as so much 'gobbledygook'. This was a puzzling and quite gratuitous rude noise because no informed man questions its truth.

The doctrine states that all the separate activities of scientists, however disparate their immediate purposes may appear to be, mutually support and sustain each other. For this and many other reasons, the distinction between pure and applied science cannot be a categorical one. However, for the sake of fairness it must be admitted that the champions of pure learning have in one respect brought today's mercenary reappraisal of their activities upon themselves: they have attempted to justify academic science by calling attention to the useful or financially profitable advances that have grown unpredictably out of their activities in the past. If they themselves are prepared to evaluate their work by a scale calibrated in dollars, they should not resent it if others do the same.

As with all expenditure ultimately subsidized by the Government, it boils down to a matter of priorities; and when their attention was called to it, many scientists in England expressed puzzlement at a scale of values which can allot something between seven and eight million pounds a year for the maintenance of military bands in the regular forces and the reserve. There is also a misunderstanding of the nature of the research process. It is because they don't know how to, not because they don't want to, that research workers don't go straight to the heart of cancer and rheumatoid arthritis without spending time and money on fundamental research.

The system by which practically valuable results grow in some-times unexpected ways from wide-ranging exploratory research is not a very satisfactory one, and no one thinks it is. If Lord Rothschild, or anybody else, can devise a better methodology, we research workers will adopt it straightaway.

17

Is the Scientific Paper a Fraud?
(1963)

I have chosen for my title a question: Is the scientific paper a fraud? I ought to explain that a scientific 'paper' is a printed communication to a learned journal, and scientists make their work known almost wholly through papers and not through books, so papers are very important in scientific communication. As to what I mean by asking 'is the scientific paper a fraud?'—I do not of course mean 'does the scientific paper misrepresent facts', and I do not mean that the interpretations you find in a scientific paper are wrong or deliberately mistaken. I mean the scientific paper may be a fraud because it misrepresents the processes of thought that accompanied or gave rise to the work that is described in the paper. That is the question, and I will say right away that my answer to it is 'yes'. The scientific paper in its orthodox form does embody a totally mistaken conception, even a travesty, of the nature of scientific thought.

Just consider for a moment the traditional form of a scientific paper (incidentally, it is a form which editors themselves often insist upon). The structure of a scientific paper in the biological sciences is something like this. First, there is a section called the 'introduction' in which you merely describe the general field in which your scientific talents are going to be exercised, followed by a section called 'previous work' in which you concede, more or less graciously, that others have dimly groped towards the fundamental truths that you are now about to expound. Then a section on 'methods'—that is OK. Then comes the section called 'results'. The section called 'results' consists of a stream of factual information in which it is considered extremely bad form to discuss the significance of the results you are getting. You have to pretend that your mind is, so to speak, a virgin receptacle, an empty vessel, for information which floods into it from the external world for no reason which you yourself have revealed. You reserve all appraisal of the scientific evidence until the 'discussion' section, and in the discussion you adopt the ludicrous pretence of asking yourself if the information you have collected actually means anything; of asking yourself if any general truths are going to emerge from the contemplation of all the evidence you brandished in the section called 'results'.

Of course, what I am saying is rather an exaggeration, but there is more than a mere element of truth in it. The conception underlying this style of scientific writing is that scientific discovery is an inductive process. What induction implies in its cruder form is roughly speaking this: scientific discovery, or the formulation of scientific theory, starts with the unvarnished and unembroidered evidence of the senses. It starts with simple observation—simple, unbiased, unprejudiced, naïve, or innocent observation—and out of this sensory evidence, embodied in the form of simple propositions or declarations of fact, generalizations will grow up and take shape, almost as if some process of crystallization or condensation were taking place. Out of a disorderly array of facts, an orderly theory, an orderly general statement, will somehow emerge. This conception of scientific discovery in which the initiative comes from the unembroidered evidence of the senses was mainly the work of a great and wise, but in this context, I think, very mistaken man—John Stuart Mill.

John Stuart Mill saw, as of course a great many others had seen before him, including Bacon, that deduction in itself is quite powerless as a method of scientific discovery—and for this simple reason: that the process of deduction as such only uncovers, brings out into the open, makes explicit, information that is already present in the axioms or premises from which the process of deduction started. The process of deduction reveals nothing to us except what the infirmity of our own minds has so far concealed from us. It was Mill's belief that induction was the method of science—'that great mental operation', he called it, 'the operation of discovering and proving general propositions'. And round this conception there grew up an inductive logic, of which the business was 'to provide rules to which, if inductive arguments conform, those arguments are conclusive'. Now John Stuart Mill's deeper motive in working out what he conceived to be the essential method of science was to apply that method to the solution of sociological problems: he wanted to apply to sociology the methods which the practice of science had shown to be immensely powerful and exact.

It is ironical that the application to sociology of the inductive method, more or less in the form in which Mill himself conceived it, should have been an almost entirely fruitless one. The simplest application of the Millsian process of induction to sociology came in a rather strange movement called Mass Observation. The belief underlying Mass Observation was apparently this: that if one could only record and set down the actual raw facts about what people do and what people say in pubs, in trains, when they make love to each other, when they are playing games, and so on, then somehow, from this wealth of information, a great generalization would inevitably emerge. Well, in point of fact, nothing important emerged from this approach, unless somebody has been holding out on me. I believe the pioneers of Mass Observation were ornithologists. Certainly they were man-watching—were applying to sociology the very methods which had done so much to bring ornithology into disrepute.

The theory underlying the inductive method cannot be sustained. Let me give three good reasons why not. In the first place, the starting point of induction, naïve observation, innocent observation, is a mere philosophic fiction. There is no such thing as unprejudiced observation. Every act of observation we make is biased. What we see or otherwise sense is a function of what we have seen or sensed in the past.

The second point is this. Scientific discovery or the formulation of the scientific idea on the one hand, and demonstration or proof on the other hand, are two entirely different notions, and Mill confused them. Mill said that induction was the 'operation of discovering and proving general propositions', as if one act of mind would do for both. Now discovery and proof could depend on the same act of mind, and in deduction they do. When we indulge in the process of deduction—as in deducing a theorem from Euclidian axioms or postulates—the theorem contains the discovery (or, more exactly, the uncovery of something which was there in the axioms and postulates, though it was not actually evident) and the process of deduction itself, if it has been carried out correctly, is also the proof that the 'discovery' is valid, is logically correct. So in the process of deduction, discovery and proof can depend on the same process. But in scientific activity they are not the same thing—they are, in fact, totally separate acts of mind.

But the most fundamental objection is this. It simply is not logically possible to arrive with certainty at any generalization containing more information that the sum of the particular statements upon which that generalization was founded, out of which it was woven. How could a mere act of mind lead to the discovery of new information? It would violate a law as fundamental as the law of conservation of matter: it would violate the law of conservation of information.

In view of all these objections, it is hardly surprising that Bertrand Russell in a famous footnote that occurs in his *Principles of Mathematics* of 1903 should have said that, so far as he could see, induction was a mere method of making plausible guesses. And our greatest modern authority on the nature of scientific method, Professor Karl Popper, has no use for induction at all: he regards the inductive process of thought as a myth. 'There is no need even to mention induction,' he says in his great treatise, on *The Logic of Scientific Discovery*—though of course he does.

Now let me go back to the scientific papers. What is wrong with the traditional form of scientific paper is simply this: that all scientific work of an experimental or exploratory character starts with some expectation about the outcome of the enquiry. This expectation one starts with, this hypothesis one formulates, provides the initiative and incentive for the enquiry and governs its actual form. It is in the light of this expectation that some observations are held relevant and others not; that some methods are chosen, others discarded; that some experiments are done rather than others. It is only in the light of this prior expectation that the activities the scientist reports in his scientific papers really have any meaning at all.

Hypotheses arise by guesswork. That is to put it in its crudest form. I should say rather that they arise by inspiration; but in any event they arise by processes that form part of the subject-matter of psychology and certainly not of logic, for there is no logically rigorous method for devising hypotheses. It is a vulgar error, often committed, to speak of 'deducing' hypotheses. Indeed one does not deduce hypotheses: hypotheses are what one deduces things from. So the actual formulation of a hypothesis is—let us say a guess; is inspirational in character. But hypotheses can be tested rigorously— they are tested by experiment, using the word 'experiment' in a rather general sense to mean an act performed to test a hypothesis, that is, to test the deductive consequences of a hypothesis. If one formulates a hypothesis, one can deduce from it certain consequences which are predictions or declarations about what will, or will not, be the case. If these predictions and declarations are mistaken, then the hypothesis must be discarded, or at least modified. If, on the other hand, the predictions turn out correct, then the hypothesis has stood up to trial, and remains on probation as before. This formulation illustrates very well, I think, the distinction between on the one hand the discovery or formulation of a scientific idea or generalization, which is to a greater or lesser degree an imaginative or inspirational act, and on the other hand the proof, or rather the testing of a hypothesis, which is indeed a strictly logical and rigorous process, based upon deductive arguments.

This alternative interpretation of the nature of the scientific process, of the nature of scientific method, is sometimes called the hypothetico-deductive interpretation and this is the view which Professor Karl Popper in *The Logic of Scientific Discovery* has persuaded us is the correct one. To give credit where credit is surely due, it is proper to say that the first professional scientist to express a fully reasoned opinion upon the way scientists actually think when they come upon their scientific discoveries—namely William Whewell, a geologist, and incidentally the Master of Trinity College, Cambridge—was also the first person to formulate this hypothetico-deductive interpretation of scientific activity. Whewell, like his contemporary Mill, wrote at great length—unnecessarily great length, one is nowadays inclined to think—and I cannot recapitulate his argument, but one or two quotations will make the gist of his thought clear. He said: 'An art of discovery is not possible. We can give no rules for the pursuit of truth which should be universally and peremptorily applicable.' And of hypotheses, he said, with great daring—why it was daring I will explain in just a second—'a facility in devising hypotheses, so far from being a fault in the intellectual character of a discoverer, is a faculty indispensable to his task'. I said this was daring because the word 'hypothesis' and the conception it stood for was still in Whewell's day a rather discreditable one. Hypotheses had a flavour about them of what was wanton and irresponsible. The great Newton, you remember, had frowned upon hypotheses. 'Hypotheses non fingo', he said, and there is another version in which he says 'hypotheses non sequor'—I do not pursue hypotheses.

So to go back once again to the scientific paper: the scientific paper is a fraud in the sense that it does give a totally misleading narrative of the processes of thought that go into the making of scientific discoveries. The inductive format of the scientific paper should be discarded. The discussion which in the traditional scientific paper goes last should surely come at the beginning. The scientific facts and scientific acts should follow the discussion, and scientists should not be ashamed to admit, as many of them apparently *are* ashamed to admit, that hypotheses appear in their minds along uncharted byways of thought; that they are imaginative and inspirational in character; that they are indeed adventures of the mind. What, after all, is the good of scientists reproaching others for their neglect of, or indifference to, the scientific style of thinking they set such great store by, if their own writings show that they themselves have no clear understanding of it?

Anyhow, I am practising what I preach. What I have said about the nature of scientific discovery you can regard as being itself a hypothesis, and the hypothesis comes where I think it should be, namely, it comes at the beginning of the series. Later speakers will provide the facts which will enable you to test and appraise this hypothesis, and I think you will find—I hope you will find—that the evidence they will produce about the nature of scientific discovery will bear me out.

18

The Pissing Evile
(1983)

The discovery of insulin may be rated the first great triumph of medical science. The first important contribution of the great pharmaceutical companies to human welfare was surely the preparation, purification, standardization, and marketing of insulin in a form suitable for self-administration by the afflicted patients.

The entire episode brought to an end, with an appropriately reverberant thunderclap, the long epoch of therapeutic nihilism described by Lewis Thomas in *The Youngest Science*.* The insulin story begins, of course, as other medical stories begin, at the bedside—with the taking of a history and an appraisal of the patient's general health. The history would be loss of weight, debility, and general malaise, of intractable thirst, the continual passing of urine that led to a seventeenth-century London surgeon's describing diabetes as the 'pissing evile', and apparent susceptibility to infections. All this would raise in the physician's mind a suspicion of *diabetes mellitus*—sugar diabetes—soon to be directly confirmed by testing the urine or watching house-flies congregate around a drop of evaporated urine. Older physicians still recount these diagnostic exercises in order to rebuke or silence enthusiastic young medical scientists who babble incoherently about the place of nuclear magnetic resonance spectroscopy in a country practice. The discovery and marketing of insulin put it for the first time within the power of the profession to restore to something like normal life victims of juvenile diabetes who would otherwise have had before them a life of invalidism terminated by early death.

* See Chapter 13.

The insulin story is here recounted as a sabbatical exercise by the Professor of Canadian History in the University of Toronto.* It is not now likely that there will be any further windfall of evidence relating to the matter: this, it seems, will be the definitive history. More than that, it is well written and highly readable.

Before the story begins, physiological research had already made it clear that the pancreas had something to do with diabetes, very probably by the manufacture of an internal secretion. The extraction of such a secretion—it was later discovered to be a protein—was felt to be a chancy and difficult business because the secretion might be destroyed during extraction by the powerful digestive enzymes already known to be manufactured by the pancreas. A young surgeon, Frederick Banting, of London, Ontario, conceived the idea that extraction of a hypothetical secretion would be made easier by a ligation of the pancreatic duct to bring about atrophy of the enzyme-secreting parts of the pancreas, which could thereby be converted into a predominantly endocrine organ: that is, into an organ principally responsible for manufacturing internal secretions, which would then be liberated directly into the bloodstream instead of travelling through a duct to their place of action. The Professor of Physiology at the University of Toronto, J. J. R. Macleod, was not carried away by the notion, but he owes his place in history to having thought well enough of it to give Banting a room and the services of a medical student, Charles Best, whose principal duty would be to carry out the crucially important job of estimating the concentration of reducing sugars in blood and urine—a simple procedure now that it has been streamlined and virtually automated by the pharmaceutical companies' diagnostic kits, but something which at the time required single-minded attention.

Although the research of Banting and Best, like all other research enterprises everywhere else, encountered some disheartening snags to begin with, they had what experimenters call a reliable 'system' to work with. Dogs deprived of their pancreases live no longer than a few weeks: if the pancreatic extracts were to work, they would bring the blood-sugar concentration clattering down and allow these dogs to survive for more than a couple of weeks.

* Michael Bliss, *The Discovery of Insulin* (Paul Harris, 1983).

In the fall of 1921 Professor Macleod returned from holiday to find that Banting and Best had kept a diabetic dog, Marjorie, alive for twenty days by means of pancreatic extracts. Macleod now put the whole laboratory to work on the problem and wisely called in his colleague Dr J. B. Collip and a big American pharmaceutical company to help with the extraction and large-scale production. The drug company, Eli Lilly, brought with them, not only great practical knowhow, but whatever impetus might be imported to the research by the prospect both of huge profits and of conferring a great benefaction on mankind. Considering that the Toronto team did not know the commercial practices of pharmaceutical companies as they became apparent from their skulduggery in the development of penicillin, it was remarkably prescient of the University of Toronto to patent the extraction process, so far as they were able to do so, in order to exercise some degree of control over the quality of the manufactured product. Meanwhile Banting and Best were ready to attempt a clinical trial of their product on a boy (Leonard Thompson) at the Toronto General Hospital. One of the first people in England to benefit from insulin treatment was a young physician at King's College Hospital, R. D. Lawrence, who survived to become England's principal authority on diabetes.

The experiments carried out by the Toronto team were, as sometimes happens when great discoveries are announced, the subject of sour and unfair disparagement—in this case, by a Dr Ffrangcon Roberts, who in turn was sharply rebuked by Sir Henry Dale OM, FRS.

Some very great advances in medicine have been brought about by clinical reasoning alone, without recourse to experimentation on animals—vaccination against smallpox was one such discovery—but this is the exception rather than the rule. Experimentation, however, is at all times unconditionally necessary. The tests of the safety and therapeutic efficacy of medical procedures are carried out either on the poor, as Bernard Shaw implied in the uproariously funny preface to *The Doctor's Dilemma*, or upon prisoners, for as Voltaire records in his letters from England, the efficacy and safety of variola-tion against smallpox was carried out with the enthusiastic conni-vance of King and Court on six condemned felons in Newgate Gaol.

It is better that laboratory animals should be used than that tests should be made directly upon human beings. So far as insulin is concerned, it was only by experimentation on dogs that it came to be learnt that removal of something manufactured by the pancreas caused diabetes. At one time anti-vivisectionists—who perform a very useful function in safeguarding animals against wanton experimentation inspired more by curiosity than by serious medical or scientific intentions—tried to belittle the discovery of insulin by claiming that the death-rate from diabetes did not decline significantly for many years after the introduction of insulin into clinical practice. This, however, was due to an accident of medical demography: whatever may have been the proximate cause of death, diabetes was always entered as a contributory cause on the patient's death certificate if he had at one time suffered from it. In the continuing debate between experimentalists and champions of the rights of animals, the discovery of insulin remains a shining example of the benefactions experimental animals have conferred upon man.

Research upon diabetes is still in progress. Scientific research has now demonstrated what was until recently only a clinical surmise— that the insulin-dependent diabetes of juveniles is a quite different clinical entity from the diabetes that can be controlled by diet which sometimes occurs in middle age. The suspicion is growing, moreover, that there is an important auto-immune element in diabetes: an element of immunological self-destruction, perhaps of viral origin though this is not yet certain. If this auto-immunity turns out to be an important factor in the disease it may be that control of auto-immunity, such as is attempted in the treatment of rheumatoid arthritis, thyroiditis, and ulcerative colitis, will play an important part in the treatment of diabetes. By far the most exciting future possibility, however, is that we will be able to transplant insulin-forming tissue, either in the form of a whole pancreas or in the form of populations of individual cells separated from islet tissue—the tissue which manufactures insulin.

'This is a book about life, disease, death, salvation, and immortality.' It is also about clinical appearances and earlier treatments of diabetes— amongst which Osler frowningly noted the use of blistering and the administration of opiates. Unhappiest of all the treatments for diabetes propounded in these dark days of medicine was that of a French doctor, Piorry, who sought to make good the wasting associated with diabetes by supplementing the patients' carbohydrate intake with extra large quantities of sugar. At a time when these, as we now see, idiotic notions were still prevalent, Paris was besieged by the German Army and in the semi-starvation consequent upon the state of siege, a French doctor made amends for Piorry's aetiological blunder by noting the disappearance of glycosuria from diabetic patients in extreme privation. This doctor, Bouchardat, also noted that exercise increased a diabetic's tolerance of carbohydrate.

Bliss, who has a nice sense of what is important and what is not, describes how it comes about that the metabolism of fat in the extremities of diabetic starvation imparts a sweetish odour to the breath—the 'ketosis' which I remember J. B. S. Haldane's unhesitatingly proposing as the cause of that 'odour of sanctity' which surrounded holy men who starved themselves for the greater glory of God.

The first salvo in the struggle against diabetes was fired in 1889 by experiments in which Oscar Minkowsky and Josef von Mehring showed that pancreatectomy of dogs led to all the symptoms of diabetes. We learn thereafter how it had become almost certain by the time of Banting and Best that certain cells in the pancreas, those of the 'islets of Langerhans', manufactured an internal secretion the deprivation of which would lead to diabetes. It was quite new to me that just about the time that Banting and Best started their work, a Romanian scientist, Nicholas Paulesco, started to treat diabetes with a pancreatic extract called 'pancreine'. These pioneering efforts were forgotten—crushed by the resources and enormous know-how of the goal-oriented research that happened in Toronto and the USA. Paulesco's claim was indeed weighty enough to have induced the International Diabetes Federation to publish in 1971 a report which, while acknowledging Paulesco's priority, attributes the principal credit for the introduction of insulin therapy to the research of Banting and Best. Bliss records that Charles Best read Paulesco's publication and indexed it in his file. Best seemed however, to have misunderstood Paulesco's paper, to which he made an inaccurate reference in his own first publication of 1922. When Paulesco came to hear of Banting's work he wrote to ask him for reprints, but Banting, a bad correspondent anyway, did not reply. On the award of the Nobel Prize for the discovery of insulin Paulesco's appeal to the Nobel Committee was ignored. It is a sad story of the triumph of the big battalions over lonely amateurs and must be repeated much more often than we are aware of.

After Bliss's account of these early contributions to the understanding of diabetes, he turns naturally to the life and contribution of Dr Frederick Grant Banting, who enrolled as a medical student in 1912 in one of North America's largest medical schools—in the University of Toronto on the shores of Lake Ontario. Bliss says he was a tall and handsome young man who spent much of his spare time with his girlfriend. He was intended for the Methodist ministry and perhaps for this reason never learned to dance. War shortened the medical course and in 1917, after his fifth year, he left for England as a member of the Canadian Army Medical Corps. 'He saw a fair bit of action and received the Military Cross for his courage under fire at Cambrai, where he was wounded in the arm by shrapnel.' By September 1919 he returned as a surgeon to the Hospital for Sick Children in Toronto, where he had been profoundly influenced by a brilliant chief surgeon, Clarence Starr.

Banting set up in practice in London, Ontario, but patients wouldn't come and in one month he earned only $4: 'every week of medical practice drove him deeper in debt'. Banting built a garage where he dabbled in oils and tinkered with his rotten car but he went on studying for his FRCS and got a part-time job as a demonstrator in surgery and anatomy in the Western University in London, Ontario. Although he lectured the physiology students on carbohydrate metabolism, he had no special interest in diabetes and no practical experience of it—indeed, he failed to recognize it in a friend and classmate.

While preparing his lecture on carbohydrates Banting read a paper in the journal *Surgery, Gynaecology and Obstetrics*, on the obstruction of the pancreatic duct by a stone, which had apparently led to atrophy of the cells secreting digestive juices, while the cells of the islets of Langerhans, already reputed to manufacture the antidiabetic hormone, remained intact. 'The sole importance of Barron's

article in the history of medicine is that Fred Banting happened
to read it in the evening of a day he had been thinking about
carbohydrate metabolism.'

In his personal memoir of 1940, *The Story of Insulin*, Banting
describes the sleepless night he spent thinking constantly of Barron's
paper and wondering if ligation of the pancreatic duct might not
make it possible to extract an anti-diabetic principle from the pancreas
free of its enzymic secretion. He got up and wrote down his idea in a
little memo at 2 a.m. on 31 October 1920—this, Bliss notes, dispels
the myth 'that the idea came to him in a dream'. Many scientists will
recognize the fretful churning of Banting's mind as something quite
common in the generation of scientific ideas. Banting's notebook,
now in the archives of the Academy of Medicine in Toronto,
contains a very explicit programme of research. (The authenticity of
these notes is endorsed by their misspelling of two key words.)
Banting soon sought an interview with the Professor of Physiology
in the University of Toronto, Macleod, to whom he explained his
notion, asking for facilities to try it out in Macleod's laboratory.
Macleod offered his department's hospitality, but it was a very big
step for Banting to give up his practice and medical school job, so
there was quite a long period of indecision and prevarication.
Banting wrote later on to Macleod specifying a two and a half month
period in 1921 as the time when he would like to take up Macleod's
offer, if it still stood. It did, and Banting, initially with Macleod's
surgical assistance, embarked on his programme with the least
possible delay.

His scheme—a perfectly rational one—was to execute pancreatectomy in two stages. The first involved ligating the pancreatic
ducts that carry the digestive juice and freeing the pancreas from its
other bodily attachments, leaving a small piece intact and well
vascularized in order to protect the dog from diabetes. It was hoped
that after an interval the cells manufacturing digestive juices would
wither and make it easier to extract the anti-diabetic principle
without running the risk of its being degraded by digestive enzymes.
Such extracts might then be injected into dogs made diabetic by
completing the two-stage pancreatic removal. Macleod gave not
only surgical help but also very sensible advice on the extraction

procedure. Unhappily, 14 of the first 19 dogs experimented upon died. The appalling heat of a Toronto summer did not help the dogs—or the surgeons. Eventually two diabetic dogs survived and were judged suitable for experimental rescue by pancreatic extract made by crude and conventional means from the ostensibly atrophied pancreases of dogs whose pancreatic ducts had been tied off for seven weeks. The first two dogs on which they tried their procedure died comatose.

The rest is history—is *now*, I should say, after Bliss's scrupulous reconstruction from laboratory notebooks of what went on. The results from the third dog investigated were much better: pancreatic extracts brought down the blood sugar, though heated extracts and extracts made in the same way from liver and spleen did not. The dog eventually died from infection: something not to be wondered at considering that no attempt would have been made to procure sterility—there were no antibiotics and no filters such as we use today to remove bacteria. In the course of these experiments Banting had had a sharp tiff with Best about what Banting considered to be Best's slovenly laboratory procedures in carrying out his sugar estimations, but after his success with the third dog Banting wrote warmly to Macleod about Best, who, he said, 'assisted me in all the operations and taught me the chemistry so that we work together all the time and check each other's findings.'

The promise of success must have had a profound moral effect on Banting and Best, for the life they were now leading would have made a good libretto for a Puccini opera. They worked night and day, cooking steaks and eggs over a bunsen burner. Banting, who had almost no money, paid two dollars a week for a tiny cubicle in a rooming-house he had lived in as a student, and he sometimes got a meal by attending the Bible class which he had also attended as a student. The work was going splendidly; they had become friends with the dog that survived and were able to test on it a number of variations in the extraction procedure.

The Nobel Prize for the discovery of insulin was shared between Banting and Macleod. Banting shared the prize with Best and Macleod with Collip. But sharing the cash is not sharing the credit, and it was widely felt that Best had suffered an injustice, though, to be sure, his function was not ostensibly more elevated than that of an assistant. There was a lot of bitching about why Macleod was mentioned at all. He certainly did not deserve the prize for the magnitude of his scientific contribution to the great discovery. On the other hand, he did his stuff as an administrator and performed very well what he himself described as the function of an impresario. He gave Banting his chance, gave what technical advice and assistance he could, and helped to make sure that the discovery was used for the benefit of mankind and not for lining the pockets of the great pharmaceutical companies. All this deserves a place in history, but not a Nobel Prize; nor did it, on the other hand, deserve the contemptuous antipathy he received from Banting.

I know from experience that it is almost always impossible in collaborative work to allocate credit exactly and justly because of the impossibility of assessing the precise contribution of synergy, which is often the most important factor of all in the work carried out by colleagues and friends. The idea was Banting's, even if it was in some respects conceptually mistaken (ligation of the pancreatic ducts wasn't really necessary), and it is therefore appropriate that the shy, barely articulate, self-taught country doctor should be rated the great romantic hero of twentieth-century medicine.

19

Animal Experimentation in a Medical Research Institute[1]
(1972)

The Stephen Paget Lecture* has as its particular theme a defence of the use of experimental animals to enlarge medical knowledge. You may well wonder why in the year 1966 such a defence should be thought necessary, and, conversely, why the general public should demand repeated assurances that medical research is being humanely and properly conducted. I myself believe it is entirely right that the public should ask for these assurances. When I say 'properly conducted' I do not mean properly conducted *only* in respect of experiments on animals (although that happens to be the particular theme of this lecture), but properly conducted in respect of all research activities that could reasonably cause misgivings. For example, the possible dangers of clinical experimentation; the endeavour to keep people alive beyond what is thought to be their natural span by the use of medical contrivances of one kind or another—or alternatively, the morality of *not* keeping them alive when it is in principle possible to do so. Then there are the possible dangers of our great and ever-growing dependence on medical supplies and medical services, a dependence so great as to tempt people to say that one day the whole world will turn into a kind of hospital in which even the best of us will be no more than ambulatory patients; and the dangers, real or imagined, of the genetic deterioration brought about by the propagation of the genetically unfit.

[1] The National Institute for Medical Research of which I was director 1962–71.

---

* Organized by the Research Defence Society.

Some of these dangers are illusory, and can be shown to be so; but the fears and misgivings they give rise to are not illusory, and they must be allayed—by public discussion, by making the truth of these matters widely known, and by such methods as the delivery of Stephen Paget Memorial Lectures.

In saying that medical research workers should be required to give a fair account of themselves to the general public, I am talking as if the general public were a sort of all-wise body into whose care the well-being of animals could perfectly safely be entrusted. Alas—this is very far from being the case. The general public is by no means qualified to judge whether or not our human wardenship of animals is being satisfactorily discharged.

Some years ago I had the privilege of serving on a Home Office Committee 'to inquire into practices or activities which may involve cruelty to British wild mammals, whether at large or in captivity'. We took great pains to hear evidence from all interested parties, but the amount of evidence that bore on the welfare of unattractive animals, or on pests like rats, was negligible.

It is difficult not to despise the sentimental ignorance about animals that is so widely thought of as a traditional part of the British character—the kind of ignorant sentimentality that finds expression in the fatuous cry that a caged bird should be 'given its freedom'. Somebody should make the general public familiar with modern research on the dynamics of natural populations of animals: for example, the work in which Professor Lack has shown that the annual adult mortality of the European robin is as high as 60 per cent, of the starling 50 per cent, and of the sparrow no less than 45 per cent. The concern of the British public for the welfare of animals is, as a matter of fact, a rather new thing: it does not lie deep in our traditions. I think I am right in saying that the common law takes no cognizance of the rights of animals, and I do not know if it even concedes that animals can have rights. At all events, the first legislation protecting animals dates from the 1820s (the RSPCA was founded in 1824). The reason given for introducing new legislation to prohibit cock fighting was that it tended to corrupt the general public—not that it inflicted cruelty on the wretched animals themselves. Although I disapprove of pop sociology, a good case can be made for arguing that ignorant sentimentality about animals and how they live in nature grew up in proportion as people ceased to know anything about animals at first hand. The literature personalizing animals has grown up in the past hundred years. *Alice in Wonderland* was published in 1865 and *Black Beauty* in 1877, and soon the nursery came to be populated with animal familiars—Brer Rabbit and Peter Rabbit, Piglet and Donald Duck have been the conditioning stimuli of our childhood; but we must allow ourselves to grow up if we are to get any sensible conception of the nature and life of animals as they really are.

The opposite of ignorant sentimentality is humane understanding. Just how far the public has yet to go to achieve humane understanding is made very clear by the world of pet dogs and the Dog Shows.

During the past ten years or so, the British Veterinary Association— in particular the Small Animals Veterinary Association—and the Animal Health Trust have been fighting an uphill but in the main successful battle to educate dog breeders and their clients, show judges, and the 650-odd Breed Societies, into some understanding of, and some determination to cope with, the problems raised by the occurrence in many breeds of dogs of distressing or painful congenital deformities. I remember being stirred by a Presidential Address on this very theme at the Annual Meeting of the British Veterinary Association in 1954.

The Kennel Club has co-operated with the veterinary profession in the exposure and analysis of these abnormalities, and the fruits of their co-operation can be read in a series of important papers in the *Journal of Small Animal Practice*. They make sorry reading. The gist of them is that most breeds of dogs carry a cruel load of abnormalities which are the primary or secondary consequences of hereditary defects. Among them are dislocation of the kneecap or hip, gross skeletal deformities, undescended testes, deafness, retinal atrophy, chronic dermatitis, ingrowing eyelashes, and chronic respiratory distress; they extend even to hyperexcitability, mental deficiency, or downright idiocy.

Now these deformities are of two kinds. Some are quite unwanted, and are indeed accidental. They have been unluckily fixed by in-breeding, and they remain in the stocks because breeders are more anxious to sell and unwilling to cull. These abnormalities are not approved of, but they are condoned. Other congenital abnormalities are deliberately bred for; they are show points; they are among the defining characters of the breed. I do not understand how anyone of educated sensibility can admire the bow legs and poor crumpled face of the bulldog, the spinal deformity that gives him his gay twirly tail, the palatal abnormalities that make it so difficult for him to breathe. No one with a real understanding of animals could applaud a show stance made possible by a congenital dislocation of the hip. We should all applaud the British Veterinary Association and the Animal Health Trust for the stand they have taken; and let me add that their criticism of breeders and of show judges was very much more warmly expressed than my own.

The welfare of animals must depend on an *understanding* of animals, and one does not come by this understanding intuitively: it must be learned. I once knew a little girl, who having been told that frogs were rather engaging creatures, befriended a frog. Her first thought was that it needed warming up, because it felt so cold. Her first lesson in the humane understanding of animals was that frogs do not like being warmed up and do better at the temperature of their environment.

Fortunately, some humane and learned organizations do exist to promote the welfare of animals and to educate the public to understand animals as they really are, so that they need no longer rely on some supposedly intuitive understanding of what animals think or feel. One of the most important of these organizations is UFAW, the Universities Federation for Animal Welfare. I had the pleasure of being the Chairman of its Scientific Advisory Committee for some few years. Among these organizations I include that department of the Home Office which authorizes and supervises all experimentation on animals in Institutes such as my own. Not everyone realizes what a high proportion of medical research workers in this country do warmly approve of the restriction of animal experimentation to people who are qualified to carry it out. This is not to say that the existing Home Office regulations are flawless or that certain administrative changes in their working are not now widely thought to be desirable.

After this long preamble—I intended it to be so—let me now say something about the care of and the use of animals in the National Institute for Medical Research, the largest research institute of its kind in the Commonwealth.

The NIMR is a sort of microcosm of basic medical research.

So far as is possible or practicable, the animals used in the Institute are bred within its precincts. We like to think of ourselves as the pioneers in this country of the careful and the scientifically informed husbandry of laboratory animals. The animals are in the charge of a scientific division of the Institute headed by a senior veterinary scientist. The Superintendent of the Division is responsible not only for its day-to-day running, but also for the training of animal technicians—educating them for a career which, thanks to Mr D. J. Short's* efforts as much as to anybody's, now offers the prospect of a rewarding and interesting life in what has come to be thought of as a profession ancillary to medical science. The establishment of animal technicians as a recognized profession and the regulation of examination standards by an Institute were projects in which the National Institute for Medical Research is proud to have played a leading part.

This was a revolution, for in the old days the care of animals was too often entrusted to kindly and well meaning, but often not very bright, old men. There have been two other such revolutions in laboratory animal husbandry. The second was the provision and use of animals of known genetic composition and history—notably of inbred animals and of first generation hybrids between inbred strains. This innovation met with a good deal of opposition from the medical profession. It was contended that pure bred animals were in some way artificial and unnatural, and that the results secured by using them would be misleading or unrepresentative. This criticism is, of course, based on a misunderstanding of the nature and purposes of medical research; one might with equal justice reproach the chemist for basing his researches on the use of pure compounds.

The third revolution, which is still in progress, is the control of

* Head of the Animal Division.

infectious disease. The animals in Research Institutes are of necessity kept at a population density which makes them an easy prey to epidemics. To control these epidemics—or rather, to prevent their occurrence in the first instance—is essentially a problem in medical or sanitary engineering. The same is true of the control of epidemics in human populations. Indeed, the actuarial status of the experimental animals bred in almost all laboratories today is still too much like that of a human population in the sixteenth century: a grievously high proportion of the animals still die from intercurrent infection, and without the consolation of believing that they go to heaven. The principle of protecting animal colonies from the attacks of pathogenic organisms had also to be fought for, but the battle has been won, and within the next few years all major biological Research Institutions will re-found their animal colonies on what is called a 'specific pathogen free' (SPF) basis.

In the National Institute for Medical Research, as in the country generally, the largest single users of experimental animals are those responsible for the standardization and safety control of drugs and vaccines. The National Institute is in fact an agent of the World Health Organization for defining international standards of those drugs and biological products which can only be assayed and tested by biological methods; and we are also agents of the Ministry of Health for checking the safety of vaccines and other agents of an immunological nature used in general medical practice. The two scientific divisions of the Institute responsible for this work make use of about 25,000 experimental animals a year; the control of polio vaccines makes use of some 2,000 monkeys a year. It is a large number in any absolute sense, but very small in proportion to the number of children at risk. Of course, no one is satisfied with the use of experimental animals for these purposes. The most determined efforts are constantly being made to find substitutes for the use of animals in standardization and control. To describe the directions that research is taking, I cannot improve on Dr W. M. S. Russell's '3 Rs' of humane laboratory practice; Reduction, Refinement, and Replacement. The number of animals used may be reduced only by increasing the amount of information to be secured from the study of any one. One method of doing so is to use genetically standardized animals—not necessarily inbreds. 'Refinement' is essentially a matter of increasing the precision of individual assays. Our ultimate goal, however, is the replacement of animals altogether. For drugs, we look forward ultimately to chemical assays, or at least to the adoption of *in vitro* methods, such as the immunological assays now being developed for the standardization of protein hormones. For the safety control of virus vaccines, e.g. polio vaccine, everyone hopes that the cytopathic changes produced in cells in tissue cultures will prove to be sufficiently discriminating and reliable. Slowly but progressively all these ambitions are being achieved.

These activities of the Institute are services, though they are services underpinned by research. Turning now to the researches of the Institute generally, I obviously cannot describe the dozens of projects undertaken in this past year by 180 scientists supported by perhaps twice that number of qualified technicians, but I shall choose some special examples to give you some idea of their variety and range of purposes.

Some people believe that the greatest task of modern medicine is to extend to the world generally the standards of medicine and hygiene that today obtain only in the advanced industrial countries. Let me therefore first mention the Institute's researches into malaria and leprosy in collaboration with research stations in West Africa and in Malaya respectively. Malaria is still, in a numerical sense, the world's gravest disease: some two and a half million people die every year of malaria and perhaps two hundred and fifty million are afflicted by it at any one time so it is not unreasonable that less than one-thousandth of that number of animals should be used in experimental malarial reearch. Rats and monkeys are each susceptible to their own kinds of malarial infection, and the study of rats and monkeys has already taken us a long way towards elucidating the mechanism of the cyclical recurrence of malarial illness. Our understanding of leprosy is still backward, because the organism that causes it—it belongs to the same family as the tubercle bacillus—cannot yet be grown in cell-free cultures, and grows impossibly slowly when caused to infect cells in tissue culture; but rodents can be infected with their own leprosy organism, and now at last it has become possible to grow the human organism in mice. Now, for the first time, critical experimental tests on the chemotherapy of leprosy can be undertaken.

My own special research interest is in the field of transplantation, and our ambition is to overcome the immunological barriers that normally prohibit the transplantation of tissues and organs between two different human beings or, for that matter, two different mice. The transplantation of kidneys in medical practice has already enjoyed greater success than any of us dared to believe possible even as recently as five years ago. All the methods used in clinical practice to prolong the life of homografts have been founded upon experiments carried out in mice, rabbits, rats and dogs, and thanks to them, surgery will one day enter into that new dimension of accomplishment which the transplantation of organs seems to promise.

The transplantation problem is a problem in immunology. Our Institute has been described as the greatest centre of immunological research in the world, and I shall not challenge this description. A high proportion of immunological research is now directed towards inhibiting and controlling the immunological response. When that control has been achieved, as it certainly will be, its rewards will be diffused far more widely than over the domain of transplantation itself. It will become possible to relieve that huge diffuse burden of human suffering imposed upon us by the allergies, hypersensitivities, auto-immune diseases, and many other miscarriages of the immunological process.

Research at the Institute is by no means confined to the use of lower animals; perhaps no Institute makes greater use of human volunteers for those researches in which only human beings will do. The complex of viruses responsible for the common cold—viruses first defined and cultivated in the National Institute—cause their characteristic symptoms only in man. Human volunteers are therefore used at the Common Cold Research Unit, our famous outpost near Salisbury. Human volunteers are also, of necessity, used to study the adaptation of human beings to climatic stresses. The Hampstead campus of the Institute is equipped with the complex instrumentation that makes it possible to create climatic conditions even more disagreeable than those which prevail out-of-doors. One of our most distinguished human physiologists in Hampstead is studying the athletic performance of human beings at an altitude of 7,415 ft. above sea level—the height of Mexico City, where the Olympic Games were held in 1968. It would not be very informative to simulate the Olympic Games with mice.

You may think that in choosing malaria, leprosy, and transplantation as examples of the Institute's research, I am cheating—at least in the sense that I am directing your attention to research of obvious practical utility, where there is no doubting the ultimate benefit to mankind. But what about the moral credentials of so called 'pure' research—for example on the mechanisms of protein synthesis, one of our major preoccupations?

In terms of their ultimate relevance to mankind, the difference between research into protein synthesis and on malaria is a difference of immediacy, and in the degree of diffusion of their effects. The work on protein synthesis stands further from practical application than work on malaria, but its results illuminate almost the whole of biology and medicine; they illuminate normal and abnormal growth, regeneration, reproduction, the synthesis of hormones, the production of antibodies, the multiplication of viruses and bacteria—surely a big enough dividend for any scholarly investment. You may disapprove of *all* experiments using animals, but it is scientifically and medically ruinous to approve only those with obvious practical uses and to reprobate all others.

To conclude: the use of experimental animals in medical research requires justification, and I think that the general public is right to demand repeated assurances that such a justification exists. The justification lies in the advancement of human welfare, but I myself interpret 'welfare' more widely than in terms of material benefits or the conquest of disease. Human beings are so constituted that they seem temperamentally obliged to explore the world around them, to enlarge their grasp and understanding of nature. It is to this restless and insistent exploratory process that human beings owe their present place in the world. It is too late now to adopt an intellectually pastoral existence—to adopt a molluscan solution of the problems of living. To invert an epigram of Thomas Browne's, it is too late to cease to be ambitious. The use of experimental animals in laboratories to enlarge our understanding of nature is part of a far wider exploratory process, and one cannot assay its value in isolation—as if it were an activity which, if prohibited, would deprive us only of the material benefits that grow directly out of its own use. Any such prohibition of learning or confinement of the understanding would have widespread and damaging consequences; but this does not imply that we are for ever more, and in increasing numbers, to enlist animals in the scientific service of man. I think that the use of experimental animals on the present scale is a temporary episode in biological and medical history, and that its peak will be reached in ten years time, or perhaps even sooner.* In the mean time we must grapple with the paradox that nothing but research on animals will provide us with the knowledge that will make it possible for us, one day, to dispense with the use of them altogether.

* Since the middle 1960s the number of animal experiments has declined by just over a third.

20

Great Circle of Learning
(1974)

The compilation of a great encyclopaedia is such a huge undertaking that no body of men would venture upon it and see it through who were not inspired and kept going by the conviction that they were doing something new and specially good. The principles that animated the 'Society of Gentlemen in Scotland' who launched the first edition in 1771 were explained thus:

Whoever has had occasion to consult Chambers, Owen &c., or even the voluminous French *Encyclopédie*, will have discovered the folly of attempting to communicate science under various technical terms arranged in an alphabetical order. Such an attempt is repugnant to the very idea of science, which is a connected series of conclusions deducted from self-evident or previously discovered principles. It is well if a man be capable of comprehending the principles and relations of different parts of science when laid before him in one uninterrupted chain. But where is the man who can learn the principles of any science from a dictionary compiled upon the plan hitherto adopted?

The methodology was Newtonian, as befitted a work for which Newton's great discoveries created the inspiration and the need, and the sentiment was unexceptionable, for their 'new plan' was to expound the principles of science in the form of systems or distinct treatises and to explain the technical terms as they occurred in the order of the alphabet with reference to the sciences to which they belonged. This 'new plan' is that which has shaped the design of every great encyclopaedia ever since.

These principles were reaffirmed in the consciously post-Darwinian 9th edition of 1875, compiled with the advice of James Clark Maxwell and Thomas Henry Huxley, respectively the leading physical scientist and biologist of the day.

The traditional *EB* style reached its evolutionary summit in the majestic Cambridge (11th) edition of 1910–11. Whereas the first edition was avowedly a compilation or digest of contemporary works of learning, which were ransacked in order to provide copy for the entries, the 11th edition, still more than the 9th, was distinguished by the very large number of important original articles by authors who were themselves leading authorities on the subjects upon which they wrote. The 11th edition was avowedly 'dominated throughout by the historical point of view'. This characteristic, taken in conjunction with the unexampled amplitude and gravity of the work as a whole, makes the 11th edition unique among works of learning. From a scholarly—perhaps unduly donnish—point of view it will never go out of date, for whereas a conventional encyclopaedia is only as up-to-date as the facts it records are accurate, the 11th edition will remain a permanent monument to the life of the mind.

The editors and publishers of the *EB* must be fed up with being told how sadly inferior all modern *EBs* are to the great 11th edition, and they have good reason to be so, for the expansion of the Natural Sciences, the upgrowth of many Unnatural Sciences, and the demands on space made by two world wars, made it impossible to scale up a work in the style of the 11th edition to a proportionate size. Besides, most people who want to refer to an encyclopaedia do not have it in mind to worship at a shrine of learning, and those who do usually have the knowledge and the means to find out what they want to know by going direct to original sources. In spite of this defence, it must be admitted that the 14th edition in 1929 was not really a great success, for while abjuring the expansive style of the 11th edition it had acquired too much the character of a work of learning adapted to a world in which quiz contests and other such exhibitions were busily creating the illusion that braininess is to be measured in terms of the possession of factual information.

The *EB* editions in the old style had each their distinctive glories (the first had a fine piece on Fluxions, the 9th had Huxley on biology and Macaulay on Samuel Johnson; and the 11th, A. N. Whitehead on geometry); as we get to know it better the same will doubtless be found true of the 15th edition.

For the 15th edition is something very different from the 14th and it really does embody a new idea which can be regarded as a sort of evolutionary denouement of the principles upon which the *EB* was founded. Having reached a certain appropriate size, it has now, like an organism, undergone an act of fission which has divided it into three parts, each with a certain independent measure of life. The three parts are named the 'Propaedia', 'Macropaedia', and 'Micropaedia'—unhappy terms, it may be, but not out of keeping with 'encyclopaedia' itself, which is not regarded by the *OED* as an ornament to etymology.

The publishers recommend the seeker after truth to turn first to the Micropaedia, which is a conventional but outstandingly good reference work of things, names, places, topics, ideas, institutions, and technical terms—e.g., banjo, Roger Bannister, Bannockburn, backwardation, and *Banque de France* ('backwardation' is roughly speaking the opposite of 'contango'). The Micropaedia also serves as an index and is a starting-point therefore for deepening or broadening whatever knowledge it is we seek. In one respect the Micropaedia remains stoutly conventional: the inclusion of many entries bearing upon classical mythology—e.g., Sisyphus, Tantalus, Leda, and all that *galère*. In the opinion of many impatient young moderns (perhaps mostly scientists—at any rate those who have made me privy to their inmost thoughts) most classical mythology, so far from being one of the most precious possessions of our cultural heritage, giving us deep insights into human nature and the human condition, is a load of old codswallop—a term borrowed from literary criticism for the meaning of which one must turn to the *OED Supplement*, Vol. i, p. 566, for here the *EB* won't help. One day some genius of encyclopaedism will recognize this to be the case and relegate all such entries to specialized reference works. The Micropaedia is packed with illustrations, often in colour. Most of the illustrations are sensible and relevant: they are of works of art, buildings, and people's faces, etc.; but even here there are vestiges of an older tradition: e.g., the illustration of 'giraffe'.

I stake my reputation that no one who knows what a giraffe looks like found out by referring to an encyclopaedia. So widely diffused is the knowledge of what a giraffe looks like and so early in life do children seem to know, that one might be tempted under the inspiration of a great authority on psycholinguistics (A. N. Chomsky, q.v.) to suppose the knowledge genetically programmed or 'inborn'; and who would ever have thought that, in addition to being that most engaging animal *Giraffa camelopardalis*, a giraffe is also a kind of upright piano or spinet? The Encyclopaedia naturally contains much else to gratify the enquiring mind: I learn, for example, that when first devised, lawn tennis was called 'sphairistiké' (from the Greek, you know) and I have no doubt that this was the origin of the well-known expression 'Who's for sphairistiké?'

The entries in the Micropaedia are not so telegraphically written as to be disagreeable to read for entertainment and instruction; on the contrary, contributors and editors between them have settled on a relaxed but concentrated style admirably well suited to its purpose and one which makes the entries a delight to browse in (the little biographies in the Micropaedia are specially good). A bright child who grew up with access to the *EB* Micropaedia would be lucky indeed—not least for having parents who could afford it: the price of the whole work varies from £279 to £500 according to binding. For this one gets the ten-volume Micropaedia, the nineteen-volume Macropaedia, and the Propaedia. The whole weighs about 129 lbs, stretches a bit more than 4 feet and contains about 44 million words—data from which your modern computer-based literary critic can work out the mean weight and mean width of a word and other such culturally significant information. The work must be bought as a whole. I feel it is a pity that the Micropaedia cannot be bought separately, because most people only want a good sematological dictionary, and it will fulfil this purpose admirably. Moreover, as I shall explain later, the Propaedia is not indispensable.

The nineteen-volume Macropaedia ('Knowledge in Depth') is the evolutionary homologue of the vast learned treatises that were the glory of the 11th edition. The Propaedia is an introduction to the 4,000-odd articles in the Macropaedia and instead of being a measured and typographically well set out list of contents under sensible and familiar headings, it aspires to be something of a taxonomy of learning—'a kind of preamble or antechamber to the world of learning'. A slightly defensive introduction seems to me to betray an awareness that the Propaedia is not really a great success. The trouble is that a taxonomy of learning is not really very helpful unless it has a certain self-evident rightness about it. In reality the major headings are so large and so heterogeneous (e.g., 'Human Life', 'Matter and Energy', 'Human Society') and the interdigitations of learning are so profuse that the Propaedia has an arbitrary and fussy character which may put many people off. The Micropaedia is intended to be, and for most purposes *is* an adequate index: 'Cancer' was one of the topics I intended to go into fully and I found the Micropaedia entry an admirable index and guide to the matter contained in the Macropaedia.

Some of the propaedeutic articles in the Propaedia (e.g., 'The Cosmic Orphan' by Loren Eisely and 'Knowledge become Self-conscious' by Mortimer J. Adler) strike me as being too consciously literary and not sufficiently down-to-earth and gritty to make a really useful guide. By including a list of biographical articles, the Propaedia does no more than the Micropaedia does already, and this is not the only redundancy. I hope I am mistaken in my judgement of the Propaedia and that familiarity and repeated reference will prove it as useful and illuminating as the publishers believe. Nevertheless I strongly suspect that most users of the Encyclopaedia will rely upon the combined use of the Micropaedia and Macropaedia without frequent reference—perhaps even without reference at all—to the Propaedia. In adopting this approach myself, I have studied entries on a number of subjects which I rather fancy myself an authority on and a number of others of which my knowledge and understanding fall far short of what they ought to be. The entries strike me as first-rate and pitched at just the right level for general reference, i.e., neither so broadly treated as to be vacuous nor marred by fussy detail. The story line is clear where there is a story to tell.

It is not to be expected that all the literary and aesthetic opinions will be agreed with by those who read the entries. An expert on the subject tells me that it is hardly fair to describe Van Gogh's method as 'spontaneous and instinctive' rather than calculated, for Van Gogh was a highly thoughtful man who worked out very carefully how to secure the effects he intended to create.

I have not looked for minor blemishes and have indeed found very few. It can however be taken for granted that a work on this great scale will contain a large number of errors of fact or judgement, but what matters is the quality of the work considered as a whole. The new *EB* is a magnificent achievement and one that will be a source of virtually inexhaustible pleasure.

21

In Defence of Doctors
(1980)

The angle of vision from a Chair of Social Medicine such as Thomas McKeown occupied with distinction for many years in the University of Birmingham, England, is quite different from that of a physician at the bedside or a surgeon at the operating table. The difference is embodied in the following credo:

I believe that for most diseases, prevention by control of their origins is cheaper, more humane, and more effective than intervention by treatment after they occur.*

This belief, McKeown goes on to say, 'does not reduce the importance of the pastoral or samaritan role of the doctor. In some ways it increases it.' McKeown firmly repudiates the notion that his message is cognate with that which is embodied in the 'Medical Nemesis' by the author referred to in public by the late Professor Henry Miller as 'Ivan the Terrible'.†

Unfortunately the antithesis between prevention and remedy as McKeown outlines it is very seldom as simple as it might at first sight appear to be as the following examples will show.

We all know very well that the frequency of the congenital affliction known as Down's Syndrome (formerly 'Mongolism' because of Down's racist propensities) would be greatly reduced if the mean age of motherhood were also to be reduced. But some women want to have—and may for one reason or another only be able to have—a child at the age of thirty or later. Again, the work of Brian MacMahon at the Harvard School of Public Health has shown very

---

* In *The Role of Medicine* (Princeton, 1980).   † Ivan Illich.

clearly that a woman who has had her first child as a teenager stands much less risk of becoming a victim of breast cancer than a woman who has had her first child in her late twenties or *a fortiori* her thirties. This finding seems to open the door to a number of salutary preventive procedures, but in real life who is going to encourage teenagers—among them one's own daughters, perhaps—to become pregnant as soon after menarche as possible to give them extra protection in later life against a misfortune that may not befall them anyway? Prophylaxis is not enough: some women will get breast cancer no matter when their children are born, just as some people who don't smoke will get lung cancer. So no matter how energetic our preventive measures, we must still have the resources of treatment at our command.

In spite of his seniority and distinction McKeown is not above being an *enfant terrible*. The philosophic doubts which form the subject of this book

began when I went to a London Hospital as a medical student after several years of graduate research in the Departments of Biochemistry at McGill and Human Anatomy at Oxford. There were two things that struck me, almost at once. One was the absence of any real interest among clinical teachers in the origin of disease, apart from its pathological and clinical manifestations; the other was that whether the prescribed treatment was of any value to the patient was hardly noticed. . . .

Living as I do in a world of medicine and medical research I am happy to be able to affirm that from my own experience what McKeown is saying is absolute bunk.

There is a good deal more in the same vein. He says that 'there seemed to be an inverse relation between the interest of a disease to the doctor and the usefulness of its treatment to the patient'. This was why, so he tells us, 'Neurology . . . attracted some of the best minds'—and that the fascination of multiple sclerosis and amyotrophic lateral sclerosis lay in diagnostic exercises that made little difference to their progress.

Since I know many who are engaged in the treatment of or research into multiple sclerosis, and since I myself do all that is in my power to promote their work, I tentatively put forward an alternative hypothesis: the interest of multiple sclerosis is that it is a terrible disease, cruelly capricious in its incidence. It arouses perhaps more than any other the feelings of compassion that play so large a part in attracting the young into the study of medicine.

It seems to me that McKeown, who temporarily casts himself in the role of St Peter, weakens his position by resolving to admit dentists into heaven. Not one of us will deny that oral hygiene and the judicious use of fluoride are much preferable to remedial dentistry. But alas, teeth decay in spite of our best endeavours, so we still need dentists—and thank God we have them.

McKeown began to think more deeply about the problems he had just become aware of when he was appointed to a Chair of Social Medicine in the University of Birmingham, his predecessor having been G. A. Auden, father of the poet. In his Chair McKeown came to see himself 'as an academic Billy Graham who bears the glad tidings of health for the taking to a grateful people'. He formed the opinion moreover that

medical science and services are misdirected, and society's investment in health is not well used, because they rest on an erroneous assumption about the basis of human health. It is assumed that the body can be regarded as a machine whose protection from disease and its effects depends primarily on internal intervention.

When McKeown finally gets down to business after an unnecessarily discursive prolegomenon he declares it as his intention to examine 'the validity of a concept . . . on which medical activities largely rest'—the concept that the maintenance of health depends upon the understanding of the structure and function of the body and the disease processes that affect it—an approach which he regards as 'mechanistic', a word which he interprets in the sense of 'machine-like' though for many years biologists have taken it to signify 'physically determinate'. There follows some uneasy discussion of the mind/body relationship during the course of which McKeown mentions G. A. Ryle and the notion of 'Category-mistakes' without giving me the impression that he altogether understands what he is talking about.

McKeown concedes that the slow 'secular' (or long-term) improvement of human health during the nineteenth century occurred *pari passu* with the growth of our knowledge about the structure and workings of the human body. But he seems impatient with the idea that the former is a consequence of the latter, for he advocates a different view: the reduction of mortality and an improvement of health in human and animal populations are due to the greater abundance and better distribution of nutriment. This transformation annulled a principal constraint upon the growth of human and other populations, a constraint dependent on population density, namely shortage of food. Serious questions can be raised about this view as we shall see.

McKeown's chapter on 'Inheritance, Environment and Disease' has a querulous and dissatisfied air throughout: teachers before the war had urged doctors to become more keenly aware than they had been until then of the gravity and prevalence of cancer of the lung, but he chides them for having paid so little attention to aetiology and to discussing the possibility 'that the disease might be due to influences which could be modified or removed.' Things are a bit better now, though, McKeown concedes: due attention is given to the importance of smoking, exercise, and diet; moreover, conscientious clinicians, by teaching and example, try to modify the practice of their students and the behaviour of their patients. 'Nevertheless in medicine as a whole the traditional mechanistic approach remains essentially unchanged; and it will remain unchanged so long as the concept of disease is based on a physico-chemical model.'

I must say I am not clear what McKeown is complaining about. If, as is possible, cancer originates as a somatic genetic accident, this is a physical event which—if it is to be understood—must be understood in physico-chemical terms. The endeavour to understand such a phenomenon is surely not incompatible with an epidemiological analysis that might help to explain its frequency. Most sensible physicians take the view that both approaches are necessary though neither is singly sufficient.

McKeown devotes a considerable number of pages to the subject of 'Inheritance, Environment and Disease'. That genetic factors control differences of susceptibility to disease is known to be true of some diseases and not known to be false of any; I look in vain through McKeown's pages to find a statement of equal clarity and there are other ways in which this chapter disappoints. It can be inferred from McKeown's discussion on the relative influences of heredity and environment that it is not in general possible to attach any one figure to the proportional contributions of the two to differences in our character makeup (e.g., in respect of IQ scores), but although McKeown allows us to draw this inference it would surely have been better if he had explained why any such exercise is impossible. It is because the contribution made by nature to a character difference is a function of nurture (and that of nurture is a function of nature).

Further, in view of McKeown's preoccupation with preventive medicine, I had reasonably hoped for some discussion of the merits and shortcomings of the programme of J. B. S. Haldane for diminishing the frequency of 'recessive' diseases such as phenylketonuria (recessive diseases are those in which the offending gene must be inherited from *both* parents instead of—as in so-called 'dominant' diseases—from only one). The essence of the Haldane solution is the discouragement of marriage, or at all events of childbearing, by possessors of the same damaging recessive gene, a preventive measure which turns upon the fact that most victims of phenylketonuria are the offspring of a marriage between carriers of the offending gene. The shortcomings of this in many ways admirable proposal are first that carriers of recessive genes are not always identifiable and secondly that such a procedure as Haldane recommends would suspend the action of natural selection altogether and pile up still greater difficulties for future generations. Another difficulty is that putting the Haldane scheme into effect would cause a tremendous outcry from all intent upon defending the right—nay, privilege—of parents to bring into the world biochemically crippled or otherwise disadvantaged children.

The second part of McKeown's book is called 'Determinants of Health'; my spirits at once rose because McKeown is admirably well qualified to write authoritatively on the causes of the vast secular improvement in human health that has taken place over the last few centuries.

During most of man's existence it is probable that a considerable proportion of all children died or were killed within a few years of birth . . . out of ten newborn children, on average, two to three died before the first birthday, five to six by age six and about seven before maturity. In technologically advanced countries today, more than 95 percent survive to adult life.

The statistical characteristic most dramatically affected by a reduction in infantile mortality is of course the mean expectation of life at birth. McKeown does well to point out how enormously it has increased over the period during which reliable records of mortality have been kept; in Sweden it rose from between thirty and forty years in 1700 to seventy-two years for males and seventy-seven for females in 1970. Although the available records 'leave no doubt that death rates were falling from the beginning of the nineteenth century . . . there is impressive indirect evidence that the decline began somewhat earlier, probably in the first half of the eighteenth century.'

To interpret these figures McKeown says we must turn to national records of the causes of mortality which are available for England and Wales since 1838. In so far as it is possible to interpret the bills of mortality it seems that nearly 90 per cent of the total reduction of the death-rate from the beginning of the eighteenth century until today can be credited to the decline of infectious disease. The different infectious diseases contributed unequally to this decline, respiratory tuberculosis contributing most and infections of ear, pharynx, and larynx least. The standardized death-rate from smallpox in England and Wales fell from seventy-five per million in 1848–54 to two per million in 1971. The corresponding figures for scarlet fever and diphtheria were 1,016 and zero. Needless to say mortality statistics do not assess the gravity and social or personal burden of a disease; although mortality from measles is way down, it was at one time feared—for reasons subsequent research has not upheld—that multiple sclerosis, surely one of the worst of all diseases, was a late complication of measles.

The information that McKeown collates in these pages is interesting not only for those with a taste for statistical figures but also for anyone drawn to social history. Thus it is especially interesting and rather shocking to learn how great a contribution infanticide has made to infant mortality; Disraeli, McKeown tells us, 'believed that infanticide "was hardly less prevalent in England than on the banks of the Ganges" '. McKeown reminds us, too, that both criminal and legal abortion are widespread and on the increase.

I agree with McKeown's assessment of the importance of infectious diseases in human mortality. I agree also with Haldane (whose name I do not see referred to in McKeown's book) that death from infectious disease is the most important selective force that has acted upon mankind and that it has left a very nearly indelible stamp on the human genetic constitution. To give one example only: the prevalence in West Africa of the gene which converts haemoglobin A into haemoglobin S seems to be owing to the fact that those who inherit this gene from one parent only enjoy a significant degree of protection against subtertian malaria. The gene does not cause major disability except when it is inherited from both parents, when it gives rise to the grave and usually fatal blood disease known as sickle cell anaemia, causing a loss of life statistically outweighed by the gain in protection from malaria. This is an instructive example because it is important evidence for the contention that improvement of the environment—the practice of 'euphenics' as President Joshua Lederberg of Rockefeller University calls it—can lead to genetic improvement (in this case the disappearance of gene S) rather than to genetic deterioration.

The causes of the great secular decline in mortality to which I have referred is one of the great problems of social medicine. In view of the nature of his thesis, it is not surprising that McKeown should quote with approval a passage from a presidential address to the American Association of Immunologists that attributes the secular improvement in health to the establishment of a new equilibrium between infectious organisms and their victims, 'quite regardless of our therapeutic efforts. According to this interpretation,' McKeown comments, 'the trend of mortality from infectious diseases was essentially independent of both medical intervention and the vast economic and social developments of the past three centuries.

Professors of social medicine usually hold sewers in high esteem so it strikes me as surprising that McKeown evidently does not regard the institution of main sewage disposal systems as one of the 'social developments' to which he refers in the passage quoted just above. McKeown doesn't think much of antitoxins; nor, I must say, do I. I have however very little doubt of the efficacy of active immunization by toxoid substances, the evaluation of which is going to be complicated by the fact that today, persons specially at risk of contracting, say, tetanus are singled out for protective immunization and generally receive it if they have sensible and responsible physicians and employers. It is true, though, that cholera vaccine has not been proved to be efficacious. Polio vaccine, however, has been. As a medical scientist my own inclination is to give more weight to the success of the latter than the failure of the former but McKeown is determined to give the lion's share of the credit to the operation of natural selection: 'The immunological constitution of a generation is influenced largely by the mortality experience of those which precede it.'

I should be the last to depreciate the importance of natural selection and of evolutionary changes generally, but if they were a fully adequate safeguard against disease we should not get half the diseases we do. Dr David Pyke has shown, for example, that there is a clear-cut genetic element in susceptibility to the form of diabetes that presents itself in middle age or in older people. There is also a genetic element, though of a different kind, in differences of susceptibility to insulin-dependent diabetes of juvenile onset. The forces of natural selection working upon what was at one time a mortal complaint of early onset are immensely strong; but they have not been strong enough to eliminate the genetic constitution associated with a specially high susceptibility to insulin-dependent diabetes.

McKeown's views on the importance of nutrition in resistance to infectious disease are succinctly summarized as follows:

If the decline of mortality from infectious diseases was not due to a change in their character, and owed little to reduced exposure to micro-organisms before the second half of the nineteenth century or to immunization and therapy before the twentieth, the possibility that remains is that the response to infections was modified by an advance in man's health brought about by improved nutrition.

McKeown's case is founded upon the undoubted correlation between nutritional standing and susceptibility to infection, but since 'there is no direct evidence that nutrition improved in the eighteenth and early nineteenth centuries', we feel let down.

I shall use McKeown's own words to describe what he regards as evidence of improvement of food supply—his own words, lest in paraphrasing what he says I should be thought guilty of presenting an argument in such a way as to discredit him:

The most impressive evidence of the improvement in food supplies is . . . the fact that the expanded populations were fed essentially on home-grown food. The population of England and Wales increased from 5.5 million in 1702 to 8.9 in 1801 and 17.9 in 1851. Since exports and imports of food during this period were relatively small, it is clear that food production at least trebled to sustain an increase of 12.4 million in a century and a half.

The decline of mortality that occurred during the eighteenth and nineteenth centuries continued into the twentieth, but with the difference that in the twentieth century the reduction in mortality from non-infectious causes began to make an important contribution to the decline, particularly in respect of prematurity and diseases of early infancy. Deaths attributed to 'old age' diminished also, probably because improvements in diagnosis caused them now to be attributed to specific causes.

Infanticide, an important cause of death until at least the latter half of the nineteenth century, diminished during the twentieth, partly because the institution of foundling hospitals made it possible to dispose of children without killing them and partly because of the growth of contraceptive practices. The foundling hospital in St. Petersburg, McKeown reports, had 25,000 children in the mid 1830s on its rolls and admitted 5,000 annually; 30–40 per cent of the children died during the first six weeks and hardly a third reached the age of six. Those who denounce birth control procedures as morally the equivalent of murder might now pause to reflect that the reduction in the number of unwanted births has reduced the frequency of child murder—in a real, not figurative, sense.

McKeown summarizes the argument of the first and larger half of his book in terms which escape tautology only by a hair's breadth. The improvement of health that has taken place during the past three centuries was due not to what happens when we are ill, but to the fact that we do not so often become ill; and we remain well, not because of specific measures such as vaccination and immunization, but because we enjoy a higher standard of nutrition and live in a healthier environment. In at least one important respect, reproduction, we also behave more responsibly.

Turning now to the future McKeown is simplistic to a degree that takes my breath away: 'there are only two ways in which disease occurs. It results either from errors in genetic programming at fertilization, or from . . . an environment for which the genes are not adapted.' To me, a biologist, this remark is about as illuminating as to be informed that disease is caused by a departure from a state of health.

These profundities usher in passages which Jean Jacques Rousseau would surely have applauded—passages in which McKeown says that whereas genetic adaptations in response to the impact of infectious diseases may occur 'within a few generations', the requirements for health of the digestive, cardiovascular, and reproductive systems do not differ greatly from those which prevailed during man's evolution—during which we were all nomadic and had practices in respect of diet and the expenditure of energy that were profoundly changed by the agricultural revolution and the accompanying domestication of man, and were of course still more greatly changed by the coming of industry. These passages pleased me because they are evidence that even an expert on social medicine still essentially falls in with the theory of illness that prevails throughout most of the Western world; I mean the 'punishment' theory of illness, according to which illness is a judgement upon us for indolence, sloth, gluttony, or other forms of carnal self-indulgence. These are salutary reflections that reaffirm the importance of the regulation of personal conduct. A new theory of illness is now taking shape at a time when the detritus of civilization is accumulating around us: the environment gets blamed for more and more that goes amiss and it is becoming increasingly easy to blame the environment or the iniquities of *laissez-faire* capitalism rather than, as in the old days, ourselves for our medical misadventures.

When he turns to considering our health in the future McKeown seems to me again to use too broad a brush for what is in any case too large a canvas. 'Most types of mental subnormality and of congenital malformations', he writes, are the consequence of prenatal environmental influences; that, surely, is too sweeping a statement. We can only agree, though that diseases of a kind which McKeown attributes to faulty genetic programming are relatively intractable, where diseases associated with affluence are in principle preventable. A miscellaneous group of diseases is classified as potentially preventable: 'some acute respiratory infections, such as the common cold, influenza and viral pneumonia as well as gastrointestinal diseases due to viruses. More tentatively, I suggest that many psychiatric conditions are in the same class.'

In a synoptic survey of the achievements of medicine McKeown makes the familiar and important point that the death-rate from tuberculosis underwent a progressive decline that was independent of the introduction of specific remedial measures. On the other hand he is inclined to dismiss as 'perverse' Creighton's view that vaccination played virtually no part in the decline of mortality from smallpox. Since he later expresses doubts on the efficacy of medical research it is heartening to see how clear is the evidence of the beneficial effects of vaccination against poliomyelitis.

It is fully in keeping with the character of McKeown's book that he doesn't think very much of medical research and that he should quote with delighted approbation Sir Macfarlane Burnet's extraordinary *lapsus mentis* in which he said that the contribution of laboratory science to medicine had come virtually to an end. The reason he took this view, I believe, is that Macfarlane Burnet was formerly, as I was, the head of a large medical research institute devoted to 'basic' medical research and that he was as dismayed as I was at the fact that so many members of his staff were more intent upon enlarging their own reputations as 'pure scientists' than in engaging directly upon the study of medical problems.

I now think that Burnet was quite wrong and that young scientists intent upon improving natural knowledge, and Lewis Thomas, who champions them are right. As an antidote to Burnet's spiritless declaration I roundly declare that within the next ten years remedies will be found for multiple sclerosis, juvenile diabetes, and at least two forms of cancer at present considered somewhat intractable.

These remedies, moreover, will come from medical research laboratories, very likely from people ostensibly working on some quite different subject.

It is one of the sadnesses of medical education that in spite of the earnest advocacy of people in the know, ordinary medical students tend to be bored by and are even a little contemptuous of the study of social medicine and public health. In British medical schools public health is traditionally taught alongside forensic medicine, in a course compendiously known to medical students as 'rape and drains'. I should now hazard an explanation of why so many medical students depreciate the importance of social medicine because it will help to explain why McKeown's book is likely to leave so many of its readers with a feeling of uneasy dissatisfaction. Social medicine, as McKeown expounds it in his book, has to do with the illnesses and mortality of whole populations and with how they vary from time to time and from place to place. On the other hand the feelings of compassion that are thought to tempt young students into medicine soon make them realize that it is individual people, not populations, that are ill and in need of treatment. For this reason I think it likely that medical education and medical research will for many years to come remain centred upon personal rather than social medicine.

22

Son of Stroke*
(1972)

Where to be ill. Large teaching hospitals are recommended. Unless privacy is of overriding importance or you really dislike your fellow men don't go into a private ward. The nursing won't be better than in a public ward, and may easily be much worse. Besides, in a public ward you will be entertained all day by the unfolding of the human comedy and by contemplating what literary people call the Rich Tapestry of Life.

Long stays in hospital. Lying in bed for any length of time is itself a weakening process, as you will soon find when you try to get up. In adequately staffed hospitals, however, physiotherapists will keep your muscles and joints in working order.

An analogous treatment is necessary for the mind. It is a natural tendency of the mind to come to and remain at a complete standstill. This is a principle of Newtonian stature. Prolonged disuse of the brain is also bad for you. Try therefore to think or converse about something other than the exigencies of hospital life and your own piteous plight. Guests come in useful here (see below: *Visitors*) and so do books.

---

* From Sir Peter's introductory note to this light-hearted article (published in *World Medicine*, 18 Oct. 1972):

> he [the author, i.e. himself] threatened to write a book of memoirs to be called 'Stroke', hoping by this means to secure specially indulgent attention from any members of the nursing staff who wished to cut a good figure in his pages. When his attention was called to the fact that somebody has already written a book called 'Stroke', he altered his title to 'Son of Stroke'.
>
> The paragraphs that follow represent the precious distillate of this major literary work.

Books. Books, if you are well enough to read them, are crucially important for entertainment and keeping the mind in working order. Some serious works should therefore be among them. Remember, however, that if you didn't understand Chomsky when you were well, there is nothing about illness that can give you an insight into the working of his mind. Do not read a genuinely funny book within a week of having had an abdominal operation. So far from giving you stitches, it will probably deprive you of them. Books should never be so heavy as to impede the ebb and flow of the blood. A slender anthology of selected English aphorisms is strongly recommended. Ten aphorisms are normally reckoned to be equivalent to a quarter of a grain of phenobarbitone. Never take more than twenty aphorisms without medical supervision.

Sleep. If you sleep all day you must not be aggrieved if you don't sleep all night. If wakeful don't clamour for sleeping draughts, but take ten selected English aphorisms with a cup of warm milk (see above: *Books*).

Food. The food in hospitals is surprisingly good, but was not intended for people with dainty or fastidious appetites. Be warned that if you eat all day you will become disgustingly obese and thus very properly an object of derision to your friends. Desist, therefore, and give those chocolates to the nurses.

Radio. It is traditional for hospital beds to be equipped with radio outlets that don't work. Test the radio at the earliest possible opportunity, complain as soon as possible, and go on complaining until somebody does something about it. When the radio works see that kind friends bring in the *Radio Times.* Then you won't reproach yourself for missing that talk on the vegetation of Boolooland. Small transistor radios are fine, provided they have an ear monophone attachment. Otherwise they are as offensive as smoking pipes.

Sister. Your ward sister is well worth knowing and trying to make friends with, because she is almost certain to be an unusually capable and intelligent woman, which is just as well because she is a nurse, teacher, administrator, psychotherapist, and everybody's confidante. You are doing well if you manage to make friends with her.

Nurses. The qualities of character which induce young ladies to enter this overworked and underpaid profession are such as to make them specially likeable people. You will almost certainly want to do something to show your appreciation of them. Flowers and profuse gratitude are not very imaginative. It is a fact, however, that nurses are often ravenously hungry after a day's duty on the wards or soon after coming on duty after a characteristically inadequate breakfast. A secret supply of biscuits and cheese may be more acceptable and will certainly be more digestible than a pot of hothouse blooms. Another trait which nurses find agreeable is to make sure that you are visited by a stream of handsome and preferably unmarried sons, cousins, or brothers.

Visitors. Some visitors come because they love you or are genuinely concerned for you, and these you will generally welcome. Others come because they felt they ought to or to indulge their *Schadenfreude*. The latter should be got rid of as quickly as possible. This can only be done by prior arrangement with Sister who is adept at making unwanted visitors feel, as well as merely being, unwelcome.

The bodily motions. In some wards the nursing staff give the impression of regarding it as a personal affront if the entire mucosal lining of the great bowel is not evacuated daily. They attach much more importance to this than you need.

It is a rightly humiliating thought that, in spite of Man's ability to reach the moon, etc., etc., no one has yet desgined a bedpan which is not physiologically inept, uncomfortable, and somewhat obscene. The main factor in making physiotherapy supportable is the feeling that ultimately it will equip you to get out of bed yourself and look after your own needs.

Hospitality to guests: drinking. It has been said that the Middlesex Hospital will do anything for you except allow you to park in the forecourt, and in general the great teaching hospitals were erected at least half a mile from anywhere it is possible to park a car. This means that your visitors when they arrive will be harassed and exhausted and must be offered the drink which (if they have any sense) they will have brought with them. They will probably offer you a drink at the same time, but as the words 'Thanks, I don't mind if I do' rise to your lips remember the medical staff may easily mind quite a lot. They certainly will if you are suffering from a serious liver disorder. If your complaints are merely orthopaedic or mental they are not likely to object at all. But here again consult with the ward sister. Tell her, if need be, that you get a funny sort of dizzy swimming feeling in the head if you don't have a drink at six o'clock.

Serious illness: the will to live. A well-known public figure who has taken it upon himself to become the Conscience of the World has objected to transplantation as an unnatural and somewhat unwholesome method of prolonging life. But before they insist too vehemently upon the Right to Die, such people should remember that a very decided preference for remaining alive has been a major motive force with human, as with animal evolution. A very firm determination to remain alive has a mysterious therapeutic effect which helps to promote that very ambition.

The National Health Service. Don't run down the National Health Service which, in spite of faults which are inevitable in any man-made scheme, represents the most enlightened piece of social legislation of the past hundred and fifty years. If you think you can do better as a private patient attending private clinics, then good luck to you. You may need it.

23

The Life Instinct and Dignity in Dying
(1983)

Eros and Thanatos are the twin pillars that support the crumbling baroque edifice of Freudian psychoanalytic theory. Thanatos is the death instinct, which manifests itself in the aggressive and destructive activities of mind, including the wish to destroy oneself.

Psychoanalytic theory does not really achieve much support from Thanatos—the most deeply unbiological explanatory concept in Freud and I declare with the authority of fifty years as a biological teacher and researcher that there is no such thing as a death instinct; nor incidentally is there any such thing as a 'life instinct' except as a figure of speech; on the other hand the tenacity of our hold on life and the sheer strength of our preference for being alive whenever it is an option is far better evidence of a life instinct than any element of the human behavioural repertoire is evidence of a death instinct. It is odd, then, that nothing in modern medicine has aroused more criticism and resentment than the lengths to which the profession will go to prolong the life of patients who need not die if any artifice can keep them going. So have grown up the machinery and the rituals of intensive care: blood transfusion, intravenous feeding, and where necessary the mechanical ventilation of the lungs. Charity, common sense, and humanity unite to describe intensive care as a method of preserving life and not as its critics have declared, of prolonging death—a sour and ill-natured criticism which could be valid only if death was the normal consequence of the use of intensive therapy which, fortunately is by no means the case. Critics have maintained, moreover, that the mechanical contraptions and the ritual of intensive care deprive death of its dignity although the 'philosophic' opinion does not go along with this and the moral equivalent of common law (to which I shall refer below) questions the notion of dignity in death and inclines rather to the view that there is more dignity and strength in making a fight for life than in a passive abdication from what we have always thought of as our precious inheritance. Indeed, friends and relatives and onlookers sometimes feel contemptuous of a patient who doesn't try. But, as is almost always true of moral judgements in an area such as this, there are always marginal or borderline cases to which such generalizations do not apply. A patient's letting go of his hold on life rather than exposing his friends and close relations to distress and to what may be crippling expense could be a manifestation of real greatness of mind.

I suspect that all generalizations to do with dying and what the address of human beings should be to the realization of its inevitability are undermined by the uniqueness of every instance of it. Death comes in so many different ways and confronts people with so many different degrees of preparedness and so many different reasons for wanting to stay alive or wanting not to, that almost any generalization one tried to formulate could be faulted from the everyday experience of a physician.

As to the moral propriety of prolonging life, we must keep it firmly in mind that from a demographic standpoint the benefactions of medicine express themselves in increasing life expectancy. This is what virtually all medication does, including the most ordinary pills and plasters and it should be clearly understood that the difference between, on the one hand everyday medications—the plaster one might put on a cut or graze, and on the other hand a heart-lung machine—is a difference of degree only, not of principle. There is no philosophically definable dividing line between treatment that is rated dignified and morally acceptable and treatment that is declared to be an affront to the dignity of man. I have twice very nearly died as a result of cerebral vascular accidents—one a haemorrhage, the other a thrombosis—and can remember that continuing to live represented the exercise of an option which with hindsight I should rate instinctual rather than deliberate, though if I had been obliged to make a rational case for remaining alive I should not have found it difficult: I had work to do which I might have rated more important than it really was and I had a dearly beloved wife and children and moreover it was I who was the subject of the debate, and this can't—mustn't—happen to me. No thought of dignity entered my head—it is a state of mind not easily compatible with the hospital microcosm of bedpans and catheters. I needed all the help I could get to promote my ambition to remain alive. It was as allies, then, that I regarded my physicians and the apparatus of intensive care and not as so many plots to deprive me of my dignity.

How can we ever come to firm conclusions upon any of these matters? Opinions do take shape slowly though they are not ratiocinative in origin. I mean, they are not inferences from philosophic or theologic axioms or so many explicit guidances from Holy Writ: they are formed by the exercise of a sort of moral equivalent of the Common Law (the common law being that great body of received opinion about equity, justice, and the administration of law which, so far from being written down and strictly codified in statutory form is the product of a tradition based on very many generations of legal judgments and interpretations of the concepts of equity and fair dealing, all of it animated by love of fair dealing and a strong resolution to see justice done). The moral consensus originates in the same way and I believe also that the moral consensus as it makes itself known by our sense of the fitness of things is just another such heritage and in my judgement does not regard the support of life as the equivalent to prolongation of dying or as in any way an affront to human dignity or a diminishment of man.

Index

About the Author

Sir Peter Medawar, OM, FRS, born 1915 died 1987, began research in H. W. Florey's department at Oxford in the early days of the development of penicillin. His scientific reputation is based mainly on his research in immunology, which helped make transplant surgery possible. In 1960 he won the Nobel Prize for Medicine for his work on tissue transplantation. He was director of the British National Institute for Medical Research from 1962 to 1971.

Sir Peter wrote a number of books for a general audience, including *Pluto's Republic, The Limits of Science,* and with Jean Medawar, *Aristotle to Zoos.* His autobiography, *Memoir of a Thinking Radish,* was published in 1986.

The essays in this volume have been selected by his friend and executor, Dr. David Pyke, who is a registrar of the Royal College of Physicians.